## The Chec

CW00725825

# Mathematics 1
# Checkbook

**J O Bird**
BSc(Hons), FCollP, FIMA, CEng, MIEE, TEng, MIElecIE

**A J C May**
BA, CEng, MIMechE, FIElecIE, MBIM

Heinemann: London

William Heinemann Ltd
10 Upper Grosvenor Street, London W1X 9PA

LONDON    MELBOURNE    JOHANNESBURG    AUCKLAND

First published by Butterworth & Co. (Publishers) Ltd 1981
Reprinted 1983
Reprinted by William Heinemann Ltd 1986

© William Heinemann Ltd 1986

**British Library Cataloguing in Publication Data**
Bird, J. O.
    Mathematics 1 checkbook
    (The checkbooks series)
    1. Shop mathematics – Problems, exercises, etc.
    I. Title    II. May, A. J. C.
    510'246    TJ1165    80–40824

ISBN 0 434 90145 8

Printed and bound in England by
Robert Hartnoll (1985) Ltd, Bodmin

# Contents

# Note to Reader

*Checkbooks* are designed for students seeking technician or equivalent qualification through the courses of the Business and Technician Education Council (BTEC), the Scottish Technical Education Council, Australian Technical and Further Education Departments, East and West African Examinations Council and other comparable examining authorities in technical subjects.

*Checkbooks* use problems and worked examples to establish and exemplify the theory contained in technical syllabuses. *Checkbook* readers gain real understanding through seeing problems solved and through solving problems themselves. *Checkbooks* do not supplant fuller textbooks, but rather supplement them with an alternative emphasis and an ample provision of worked and unworked problems, essential data, short answer and multi-choice questions (with answers where possible).

**Level 1**
Construction Drawing
Construction Technology
Engineering Drawing
Mathematics
Microelectronic Systems
Physical Science
Physics
Workshop Processes and Materials

**Level 2**
Building Science and Materials
Chemistry
Construction Technology
Digital Techniques
Electrical and Electronic Applications
Electrical and Electronic Principles
Electronics
Engineering Drawing
Engineering Science
Manufacturing Technology
Mathematics
Microelectronic Systems
Motor Vehicle Science
Physics

**Level 3**
Building Measurement
Chemistry
Construction Technology
Digital Techniques
Electrical Principles
Electrical Science
Electronics
Engineering Design
Engineering Mathematics and Science
Engineering Science
Light Current Applications
Manufacturing Technology
Mathematics
Mechanical Science
Microelectronic Systems

**Level 4**
Building Law
Building Services and Equipment
Concrete Technology, Volumes 1 and 2
Construction Site Personnel
Construction Site Production
Construction Technology
Economics for the Construction Industry
Engineering Instrumentation and Control
Environmental Science
Mathematics

**Level 5**
Building Services and Equipment
Construction Technology

# Preface

This textbook of worked problems provides coverage of the Business and Technician Education Council Level 1 in Mathematics. However it can also be regarded as a basic textbook in mathematics for a much wider range of courses, in particular, for CSE final year students who are likely to move into engineering/science disciplines.
to move into engineering/science disciplines.

The aim of the book is to consolidate basic mathematical principles and establish a common base for further progress, taking into account the wide variety of approaches to mathematics previously encountered at earlier stages in education.

Each topic considered in the text is presented in a way that assumes in the reader little previous knowledge of that topic. This practical mathematics book contains over 300 detailed worked problems, followed by some 500 further problems with answers.

The authors would like to express their appreciation for the friendly co-operation and helpful advice given to them by the publishers. Thanks are also due to Mrs Elaine Woolley for the excellent typing of the manuscript. Finally, the authors would like to add a word of thanks to their wives, Elizabeth and Juliet, for their patience, help and encouragement during the preparation of this book.

J O Bird
A J C May
Highbury College of Technology,
Portsmouth

# 1 Revision of basic arithmetic

## A. DEFINITIONS AND OPERATIONS USED IN ARITHMETIC

1 Whole numbers are called **integers**. $+3$, $+5$, $+72$ are called **positive integers**; $-13$, $-6$, $-51$ are called **negative integers**. Between positive and negative integers is the number 0 which is neither positive nor negative.

2 The four basic arithmetic operators are: add $(+)$, subtract $(-)$, multiply $(\times)$ and divide $(\div)$.

3 For addition and subtraction, when **unlike signs** are together in a calculation, the overall sign is **negative**. Thus, adding minus 4 to 3 is $3 + -4$ and becomes $3 - 4 = -1$. **Like signs** together give an overall **positive** sign. Thus subtracting minus 4 from 3 is $3 - -4$ and becomes $3 + 4 = 7$.

4 For multiplication and division, when the numbers have **unlike signs**, the answer is **negative**, but when the numbers have **like signs** the answer is **positive**. Thus $3 \times -4 = -12$, whereas $-3 \times -4 = +12$. Similarly

$$\frac{4}{-3} = -\frac{4}{3} \quad \text{and} \quad \frac{-4}{-3} = +\frac{4}{3}$$

5 When two or more numbers are multiplied together, the individual numbers are called **factors**. Thus a factor is a number which divides into another number exactly. The **highest common factor (HCF)** is the largest number which divides into two or more numbers exactly.

6 A **multiple** is a number which contains another number an exact number of times. The smallest number which is exactly divisible by each of two or more numbers is called the **lowest common multiple, (LCM)**.

7 When a particular arithmetic operation is to be performed first, the numbers and the operator(s) are placed in brackets. Thus 3 times the result of 6 minus 2 is written as $3 \times (6 - 2)$.

8 In arithmetic operations, the order in which operations are performed are:
   (i) to determine the values of operations contained in brackets;
   (ii) multiplication and division, (the word 'of' also means multiply); and
   (iii) addition and subtraction.
   This **order of precedence** can be remembered by the word **BODMAS**, standing for Brackets, Of, Division, Multiplication, Addition and Subtraction, taken in that order.

9 The basic laws governing the use of brackets and operators are shown by the following examples:
   (i) $2 + 3 = 3 + 2$, i.e. the order of numbers when adding does not matter;
   (ii) $2 \times 3 = 3 \times 2$, i.e. the order of numbers when multiplying does not matter;
   (iii) $2 + (3 + 4) = (2 + 3) + 4$, i.e., the use of brackets when adding does not affect the result;

1

(iv) $2 \times (3 \times 4) = (2 \times 3) \times 4$, i.e. the use of brackets when multiplying does not affect the result;

(v) $2 \times (3+4) = 2(3+4) = (3+4)2 = 2 \times 3 + 2 \times 4$, i.e. a number placed outside of a bracket indicates that the whole contents of the bracket must be multiplied by that number;

(vi) $(2+3)(4+5) = (5)(9) = 45$, i.e. adjacent brackets indicate multiplication;

(vii) $2[3+(4 \times 5)] = 2[3+20] = 2 \times 23 = 46$, i.e. when an expression contains inner and outer brackets, the inner brackets are removed first.

10 When 2 is divided by 3, it may be written as $\frac{2}{3}$ or 2/3. $\frac{2}{3}$ is called a **fraction**. The number above the line, i.e. 2, is called the **numerator** and the number below the line, i.e. 3, is called the **denominator**.

11 When the value of the numerator is less than the value of the denominator, the fraction is called a **proper fraction**; thus $\frac{2}{3}$ is a proper fraction. When the value of the numerator is greater than the denominator, the fraction is called an **improper fraction**. Thus $\frac{7}{3}$ is an improper fraction and can also be expressed as a **mixed number**, that is, an integer and a proper fraction. Thus the improper fraction $\frac{7}{3}$ is equal to the mixed number $2\frac{1}{3}$.

12 When a fraction is simplified by dividing the numerator and denominator by the same number, the process is called **cancelling**. Cancelling by 0 is not permissible.

13 The **ratio** of one quantity to another is a fraction, and is the number of times one quantity is contained in another quantity **of the same kind**.

14 If one quantity is **directly proportional** to another, then as one quantity doubles, the other quantity also doubles. When a quantity is **inversely proportional** to another, then as one quantity doubles, the other quantity is halved.

15 The decimal system of numbers is based on the **digits** 0 to 9. A number such as 53.17 is called a **decimal fraction**, a **decimal point** separating the integer part, i.e. 53, from the fractional part, i.e. 0.17.

16 A number which can be expressed exactly as a decimal fraction is called a **terminating decimal** and those which cannot be expressed exactly as a decimal fraction are called **non-terminating decimals**. Thus, $\frac{3}{2} = 1.5$ is a terminating decimal, but $\frac{4}{3} = 1.33333 \ldots$ is a non-terminating decimal. $1.33333 \ldots$ can be written as $1.\dot{3}$, called 'one point-three recurring'.

17 The answer to a non-terminating decimal may be expressed in two ways, depending on the accuracy required:
   (i) correct to a number of significant figures, that is, figures which signify something; and
   (ii) correct to a number of decimal places, that is, the number of figures after the decimal point.

The last digit in the answer is unaltered if the next digit on the right is in the group of numbers 0, 1, 2, 3 or 4, but is increased by 1 if the next digit on the right is in the group of numbers 5, 6, 7, 8 or 9. Thus the non-terminating decimal $7.6183 \ldots$ becomes 7.62, correct to 3 significant figures, since the next digit on the right is 8, which is in the group of numbers 5, 6, 7, 8 or 9. Also $7.6183 \ldots$ becomes 7.618, correct to 3 decimal places, since the next digit on the right is 3, which is in the group of numbers 0, 1, 2, 3 or 4.

18 **Percentages** are used to give a common standard and are fractions having the number 100 as their denominators. For example, 25 percent means $\frac{25}{100}$, i.e. $\frac{1}{4}$, and is written 25%.

## B. WORKED PROBLEMS ON ARITHMETIC

### (a) ARITHMETIC OPERATIONS

*Problem 1* Add 27, −74, 81 and −19

This problem is written as $27-74+81-19$.

| Adding the positive integers: | 27 | Adding the negative integers: | 74 |
| --- | --- | --- | --- |
| Sum of positive integers is: | $\frac{81}{108}$ | Sum of negative integers is: | $\frac{19}{93}$ |

Taking the sum of the negative integers from the sum of the positive integers gives:

$$\begin{array}{r} 108 \\ - \phantom{0}93 \\ \hline 15 \end{array}$$   **Thus $27-74+81-19 = 15$**

*Problem 2* Subtract 89 from 123

This is written mathematically as $123-89$.   $\begin{array}{r} 123 \\ - \phantom{0}89 \\ \hline 34 \end{array}$

**Thus $123-89 = 34$**

*Problem 3* Subtract −74 from 377

This problem is written as $377- -74$. From para 3, like signs together give an overall positive sign, hence

$377- -74 = 377+74$    $\begin{array}{r} 377 \\ + \phantom{0}74 \\ \hline 451 \end{array}$

**Thus, $377- -74 = 451$**

*Problem 4* Subtract 243 from 126

The problem is $126-243$. When the second number is larger than the first, take the smaller number from the larger and make the result negative.

Thus $126-243 = -(243-126)$

$$\begin{array}{r} 243 \\ -\ 126 \\ \hline 117 \end{array}$$ **Thus $126-243 = -117$**

---

*Problem 5*  Subtract 318 from $-269$

---

$-269 - 318$. The sum of the negative integers is

$$\begin{array}{r} 269 \\ +\ 318 \\ \hline 587 \end{array}$$

**Thus, $-269 - 318 = -587$**

---

*Problem 6*  Multiply 74 by 13

---

This is written as $74 \times 13$

$$\begin{array}{r} 74 \\ 13 \\ \hline 222 \\ 740 \\ \hline 962 \end{array}$$
$\quad\longleftarrow 74 \times 3$
$\quad\longleftarrow 74 \times 10$

Adding:

**Thus $74 \times 13 = 962$**

---

*Problem 7*  Multiply 178 by $-46$

---

From para 4, when the numbers have different signs, the result will be negative. (With this in mind, the problem can now be solved by multiplying 178 by 46.)

$$\begin{array}{r} 178 \\ 46 \\ \hline 1068 \\ 7120 \\ \hline 8188 \end{array}$$   Thus $178 \times 46 = 8\ 188$ and **$178 \times (-46) = -8\ 188$**

---

*Problem 8*  Divide 1 043 by 7

---

When dividing by the numbers 1 to 12, it is usual to use a method called **short division.**

$$7\,\overline{)1\,0\overset{3}{4}\overset{6}{3}}\quad \text{quotient } 149$$

*Step 1.* 7 into 10 goes 1, remainder 3. Put 1 above the 0 of 1 043 and carry the 3 remainder to the next digit on the right, making it 34.

*Step 2.* 7 into 34 goes 4, remainder 6. Put 4 above the 4 of 1 043 and carry the 6 remainder to the next digit on the right, making it 63.

*Step 3.* 7 into 63 goes 9, remainder 0. Put 9 above the 3 of 1 043.

**Thus 1 043 ÷ 7 = 149**

*Problem 9* Divide 378 by 14

When dividing by numbers which are larger than 12, it is usual to use a method called **long division**.

```
              27
         14 ) 378
(2) 2 × 14     28
              ────
               98
(4) 7 × 14     98
              ────
               · ·
```

(1) 14 into 37 goes twice. Put 2 above the 7 of 378.

(3) { Subtract. Bring down the 8. 14 into 98 goes 7 times. Put 7 above the 8 of 378.

(5) Subtract.

**Thus 378 ÷ 14 = 27**

*Problem 10* divide 5 669 by 46

This problem may be written as $\frac{5669}{46}$ or 5669 ÷ 46 or 5669/46. Using the long division method shown in *Problem 9* gives:

```
          123
     46 ) 5669
          46
         ────
          106
           92
         ────
          149
          138
         ────
           11
```

As there are no more digits to bring down,

**5669 ÷ 46 = 123, remainder 11 or 123 $\frac{11}{46}$**

*Further problems on arithmetic operations may be found in section C, problems 1 to 21, page 16.*

(b) HIGHEST COMMON FACTORS AND LOWEST COMMON MULTIPLES

*Problem 11* Determine the HCF of the numbers 12, 30 and 42.

Each number is expressed in terms of its lowest factors. This is achieved by

5

repeatedly dividing by the prime numbers 2, 3, 5, 7, 11, 13 . . . . , (where possible) in turn. Thus

$$12 = \boxed{2} \times 2 \times \boxed{3}$$
$$30 = \boxed{2} \quad \times \boxed{3} \times 5$$
$$42 = \boxed{2} \quad \times \boxed{3} \quad \times 7$$

The factors which are common to each of the numbers are 2 in column 1 and 3 in column 3, shown by the broken lines. Hence the **HCF is 2 × 3, i.e. 6.** That is, 6 is the largest number which will divide into all of the given numbers.

*Problem 12* Determine the HCF of the numbers 30, 105, 210 and 1155.

Using the method shown in *Problem 11*:

$$30 = 2 \times \boxed{3} \times \boxed{5}$$
$$105 = \boxed{3} \times \boxed{5} \times 7$$
$$210 = 2 \times \boxed{3} \times \boxed{5} \times 7$$
$$1155 = \boxed{3} \times \boxed{5} \times 7 \times 11$$

The factors which are common to each of the numbers are 3 in column 2 and 5 in column 3.
**Hence the HCF is 3 × 5 = 15**

*Problem 13* Determine the LCM of the numbers 12, 42 and 90

The LCM is obtained by finding the lowest factors of each of the numbers, as shown in *Problems 11* and *12* above, and then selecting the largest group of any of the factors present. Thus

$$12 = \boxed{2 \times 2} \times 3$$
$$42 = 2 \quad \times 3 \qquad \times \boxed{7}$$
$$90 = 2 \quad \times \boxed{3 \times 3} \times \boxed{5}$$

The largest group of any of the factors present are shown by the broken lines and are 2 × 2 in 12, 3 × 3 in 90, 5 in 90 and 7 in 42.
   **Hence the LCM is 2 × 2 × 3 × 3 × 5 × 7 = 1260,** and is the smallest number which 12, 42 and 90 will all divide into exactly.

*Problem 14* Determine the LCM of the numbers 150, 210, 735 and 1365

Using the method shown in *Problem 13* above:

$$150 = \boxed{2} \times \boxed{3} \times \boxed{5 \times 5}$$
$$210 = 2 \times 3 \times 5 \qquad \times 7$$
$$735 = \quad 3 \times 5 \qquad \times \boxed{7 \times 7}$$
$$1365 = \quad 3 \times 5 \qquad \times 7 \qquad \times \boxed{13}$$

**The LCM is 2 × 3 × 5 × 5 × 7 × 7 × 13 = 95 550.**

*Further problems on highest common factors and lowest common multiples may be found in section C, problems 22 to 27, page 17.*

*Further problems on highest common factors and lowest common multiples may be found in section C, problems 22 to 27, page 17.*

## (c) ORDER OF PRECEDENCE AND BRACKETS

*Problem 15* Find the value of $6+4 \div (5-3)$

The order of precedence of operations is remembered by the word BODMAS (see paragraph 8).

| Thus, $6+4 \div (5-3)$ | $= 6+4 \div 2$ | (**B**rackets) |
|---|---|---|
| | $= 6+2$ | (**D**ivision) |
| | $= 8$ | (**A**ddition) |

*Problem 16* Determine the value of $13-2 \times 3+14 \div (2+5)$

| $13-2 \times 3+14 \div (2+5)$ | $= 13-2 \times 3+14 \div 7$ | (B) |
|---|---|---|
| | $= 13-2 \times 3+2$ | (D) |
| | $= 13-6+2$ | (M) |
| | $= 15-6$ | (A) |
| | $= 9$ | (S) |

*Problem 17* Evaluate $16 \div (2+6) + 18[3+4 \times 6-21]$

$16 \div (2+6) + 18[3+(4 \times 6)-21] = 16 \div (2+6) + 18[3+24-21],$
   (see para 9, (vii))
$= 16 \div 8+18 \times 6$  (B, and see para 9(v))
$= 2+18 \times 6$   (D)
$= 2+108$   (M)
$= 110$   (A)

*Problem 18* Find the value of $23-4(2 \times 7) + (144 \div 4)/(14-8)$

| $23-4(2 \times 7) + (144 \div 4)/(14-8)$ | $= 23-4 \times 14+36/6$ | (B) |
|---|---|---|
| | $= 23-4 \times 14+6$ | (D) |
| | $= 23-56+ 6$ | (M) |
| | $= 29-56$ | (A) |
| | $= -27$ | (S) |

7

*Further problems on the order of precedence and brackets may be found in section C, problems 28 to 34, page 17.*

**(d) FRACTIONS**

*Problem 19* Simplify $\frac{1}{3} + \frac{2}{7}$

The LCM of the two denominators is $3 \times 7$, i.e. 21.
Expressing each fraction so that their denominators are 21, gives:

$$\frac{1}{3} + \frac{2}{7} = \frac{1}{3} \times \frac{7}{7} + \frac{2}{7} \times \frac{3}{3} = \frac{7}{21} + \frac{6}{21}$$

$$= \frac{7+6}{21} = \frac{13}{21}$$

Alternatively:

$$\frac{1}{3} + \frac{2}{7} = \frac{\overset{\text{Step (2)}}{(7 \times 1)} + \overset{\text{Step (3)}}{(3 \times 2)}}{\underset{\text{Step (1)}}{21}}$$

*Step (1)*: the LCM of the two denominators.

*Step (2)*: for the fraction $\frac{1}{3}$, 3 into 21 goes 7 times, $7 \times$ the numerator is $7 \times 1$.

*Step (3)*: for the fraction $\frac{2}{7}$, 7 into 21 goes 3 times, $3 \times$ the numerator is $3 \times 2$.

Thus $\frac{1}{3} + \frac{2}{7} = \frac{7+6}{21} = \frac{13}{21}$, as obtained previously.

*Problem 20* Find the value of $3\frac{2}{3} - 2\frac{1}{6}$

One method is to split the mixed numbers into integers and their fractional parts. Then

$$3\frac{2}{3} - 2\frac{1}{6} = \left(3 + \frac{2}{3}\right) - \left(2 + \frac{1}{6}\right)$$

$$= 3 + \frac{2}{3} - 2 - \frac{1}{6} = 1 + \frac{2}{3} - \frac{1}{6} = 1 + \frac{4}{6} - \frac{1}{6}$$

$$= 1\frac{3}{6} = 1\frac{1}{2}$$

Another method is to express the mixed numbers as improper fractions.

Since $3 = \frac{9}{3}$, then $3\frac{2}{3} = \frac{9}{3} + \frac{2}{3} = \frac{11}{3}$

Similarly, $2\frac{1}{6} = \frac{12}{6} + \frac{1}{6} \qquad = \frac{13}{6}$

8

Thus $3\frac{2}{3} - 2\frac{1}{6}$ $= \frac{11}{3} - \frac{13}{6}$

$$= \frac{22}{6} - \frac{13}{6} = \frac{9}{6} = 1\frac{1}{2},$$

as obtained previously.

**Problem 21** Evaluate $7\frac{1}{8} - 5\frac{3}{7}$

$7\frac{1}{8} - 5\frac{3}{7} = \left(7 + \frac{1}{8}\right) - \left(5 + \frac{3}{7}\right) = 7 + \frac{1}{8} - 5 - \frac{3}{7}$

$\qquad = 2 + \frac{1}{8} - \frac{3}{7} = 2 + \frac{7 \times 1 - 8 \times 3}{56}$

$\qquad = 2 + \frac{7 - 24}{56} = 2 + \frac{-17}{56}$

$\qquad = 2 - \frac{17}{56} = \frac{112}{56} - \frac{17}{56}$

$\qquad = \frac{112 - 17}{56} = \frac{95}{56} = 1\frac{39}{56}$

**Problem 22** Determine the value of $4\frac{5}{8} - 3\frac{1}{4} + 1\frac{2}{5}$

$4\frac{5}{8} - 3\frac{1}{4} + 1\frac{2}{5} = (4 - 3 + 1) + \left(\frac{5}{8} - \frac{1}{4} + \frac{2}{5}\right)$

$\qquad = 2 + \frac{5 \times 5 - 10 \times 1 + 8 \times 2}{40}$

$\qquad = 2 + \frac{25 - 10 + 16}{40} = 2 + \frac{31}{40} = 2\frac{31}{40}$

**Problem 23** Find the value of $\frac{3}{7} \times \frac{14}{15}$

Dividing numerator and denominator by 3 gives:

$\overset{1}{\cancel{3}} \over 7 \times \frac{14}{\underset{5}{\cancel{15}}} = \frac{1 \times 14}{7 \times 5}$

Dividing numerator and denominator by 7 gives:

$\frac{1 \times \overset{2}{\cancel{14}}}{\underset{1}{\cancel{7}} \times 5} = \frac{2}{5}$

This process of dividing both the numerator and denominator of a fraction by the same factor(s) is called **cancelling**.

**Problem 24** Evaluate $1\frac{3}{5} \times 2\frac{1}{3} \times 3\frac{3}{7}$

9

Mixed numbers **must** be expressed as improper fractions before multiplication can be performed. Thus,

$$1\tfrac{3}{5} \times 2\tfrac{1}{3} \times 3\tfrac{3}{7} = \left(\tfrac{5}{5}+\tfrac{3}{5}\right) \times \left(\tfrac{6}{3}+\tfrac{1}{3}\right) \times \left(\tfrac{21}{7}+\tfrac{3}{7}\right)$$

$$= \tfrac{8}{5} \times \dfrac{\cancel{7}^{\,1}}{\cancel{3}_{\,1}} \times \dfrac{\cancel{24}^{\,8}}{\cancel{7}_{\,1}} = \dfrac{8 \times 8}{5}$$

$$= \tfrac{64}{5} = 12\tfrac{4}{5}$$

---

*Problem 25* Simplify $\dfrac{3}{7} \div \dfrac{12}{21}$

$$\dfrac{3}{7} \div \dfrac{12}{21} = \dfrac{\tfrac{3}{7}}{\tfrac{12}{21}}$$

Multiplying both numerator and denominator by the reciprocal of the denominator gives:

$$\dfrac{\tfrac{3}{7}}{\tfrac{12}{21}} = \dfrac{\dfrac{1}{\cancel{3}}{\cancel{7}} \times \dfrac{\cancel{21}^{3}}{\cancel{12}_{4}}}{\dfrac{12}{21} \times \dfrac{21}{12}} = \dfrac{\tfrac{3}{4}}{1} = \dfrac{3}{4}$$

This method can be remembered by the rule: invert the second fraction and change the operation from division to multiplication.
Thus:

$$\dfrac{3}{7} \div \dfrac{12}{21} = \dfrac{\cancel{3}^{1}}{\cancel{7}_{1}} \times \dfrac{\cancel{21}^{3}}{\cancel{12}_{4}} = \dfrac{3}{4}, \text{ as obtained previously.}$$

---

*Problem 26* Find the value of $5\tfrac{3}{5} \div 7\tfrac{1}{3}$

The mixed numbers **must** be expressed as improper fractions. Thus,

$$5\tfrac{3}{5} \div 7\tfrac{1}{3} = \tfrac{28}{5} \div \tfrac{22}{3}$$

$$= \dfrac{\cancel{28}^{14}}{5} \times \dfrac{3}{\cancel{22}_{11}} = \dfrac{42}{55}$$

---

*Problem 27* Simplify $\dfrac{1}{3} - \left(\dfrac{2}{5}+\dfrac{1}{4}\right) \div \left(\dfrac{3}{8} \times \dfrac{1}{3}\right)$

The order of precedence of operations for problems containing fractions is the

10

same as that for integers, i.e., remembered by BODMAS, (Brackets, Of, Division, Multiplication, Addition and Subtraction).    Thus,

$$\frac{1}{3} - \left(\frac{2}{5} + \frac{1}{4}\right) \div \left(\frac{3}{8} \times \frac{1}{3}\right) = \frac{1}{3} - \frac{4 \times 2 + 5 \times 1}{20} \div \frac{\cancel{3}^{\,1}}{\cancel{24}_{\,8}} \quad \text{(B)}$$

$$= \frac{1}{3} - \frac{13}{\cancel{20}_{\,5}} \times \frac{\cancel{8}^{\,2}}{1} \quad \text{(D)}$$

$$= \frac{1}{3} - \frac{26}{5} \quad \text{(M)}$$

$$= \frac{(5 \times 1) - (3 \times 26)}{15} \quad \text{(S)}$$

$$= \frac{-73}{15} = -4\frac{13}{15}$$

*Problem 28*  Determine the value of $\frac{7}{6}$ of $\left(3\frac{1}{2} - 2\frac{1}{4}\right) + 5\frac{1}{8} \div \frac{3}{16} - \frac{1}{2}$

$$\frac{7}{6} \text{ of } \left(3\frac{1}{2} - 2\frac{1}{4}\right) + 5\frac{1}{8} \div \frac{3}{16} - \frac{1}{2} = \frac{7}{6} \text{ of } 1\frac{1}{4} + \frac{41}{8} \div \frac{3}{16} - \frac{1}{2} \quad \text{(B)}$$

$$= \frac{7}{6} \times \frac{5}{4} + \frac{41}{8} \div \frac{3}{16} - \frac{1}{2} \quad \text{(O)}$$

$$= \frac{7}{6} \times \frac{5}{4} + \frac{41}{\cancel{8}_{\,1}} \times \frac{\cancel{16}^{\,2}}{3} - \frac{1}{2} \quad \text{(D)}$$

$$= \frac{35}{24} + \frac{82}{3} - \frac{1}{2} \quad \text{(M)}$$

$$= \frac{35 + 656}{24} - \frac{1}{2} \quad \text{(A)}$$

$$= \frac{691 - 12}{24} \quad \text{(S)}$$

$$= \frac{679}{24} = 28\frac{7}{24}$$

*Further problems on fractions may be found in section C, problems 35 to 47, page 17.*

## (e)  RATIO AND PROPORTION

*Problem 29*  Divide 126 in the ratio of 5 to 13

Because the ratio is to be 5 parts to 13 parts, then the total number of parts is 5 + 13, that is 18. Then,

18 parts correspond to 126

Hence 1 part corresponds to $\frac{126}{18} = 7$,  5 parts correspond to $5 \times 7 = 35$

and    13 parts correspond to $13 \times 7 = 91$

[Check: the parts must add up to the total $35 + 91 = 126 =$ the total]

11

The total number of parts is $3+7+11$, that is, 21.

Hence    21 parts correspond to 273 cm

$$1 \text{ part corresponds to } \frac{273}{21} = 13 \text{ cm}$$

$$3 \text{ parts correspond to } 3 \times 13 = 39 \text{ cm}$$

$$7 \text{ parts correspond to } 7 \times 13 = 91 \text{ cm}$$

and    11 parts correspond to $11 \times 13 = 143$ cm

i.e. **the lengths of the three pieces are 39 cm, 91 cm and 143 cm**

[Check: $39+91+143 = 273$]

Working in quantities **of the same kind**, the required ratio is $\frac{25}{425}$, i.e. $\frac{1}{17}$

That is, 25p is 1/17th of £4.25. This may be written either as:

$25 : 425 : : 1 : 17$, (stated as "25 is to 425 as 1 is to 17") or as $\frac{25}{425} = \frac{1}{17}$

The more the number of people, the more quickly the task is done, hence inverse proportion exists.

3 people complete the task in 4 hours,

1 person takes three times as long, i.e., $4 \times 3 = 12$ hours,

5 people can do it in one fifth of the time that one person takes, that is $\frac{12}{5}$ hours or **2 hours 24 minutes.**

*Further problems on ratio and proportion may be found in section C, problems 48 to 53, page 18.*

*Problem 33* Evaluate 42.7+3.04+8.7+0.06

The numbers are written so that the decimal points are under each other. Each column is added, starting from the right.

```
 42.7
  3.04
  8.7
  0.06
 54.50
```
Thus, **42.7+3.04+8.7+0.06 = 54.50**

*Problem 34* Take 81.7 from 87.23

The numbers are written with the decimal points under each other.

```
 87.23
 81.70
  5.53
```
Thus, **87.23−81.7 = 5.53**

*Problem 35* Find the value of 23.4−17.83−57.6+32.68

The sum of the positive decimal fractions is 23.4+32.68 = 56.08
The sum of the negative decimal fractions is 17.83+57.6 = 75.43
Taking the sum of the negative decimal fractions from the sum of the positive decimal fractions gives: 56.08−75.43
$$\text{i.e.} - (75.43 - 56.08) = -19.35$$

*Problem 36* Determine the value of 74.3 × 3.8

When multiplying decimal fractions: (i) the numbers are multiplied as if they are integers, and (ii) the position of the decimal point in the answer is such that there are as many digits to the right of it as the sum of the digits to the right of the decimal points of the two numbers being multiplied together.
Thus

(i)
```
   743
    38
  5944
 22290
 28234
```

(ii) As there are (1+1) = 2 digits to the right of the decimal points of the two numbers being multiplied together, (74.3 × 3.8), then
**74.3 × 3.8 = 282.34**

13

**Problem 37** Evaluate 37.81 ÷ 1.7, correct to (i) 4 significant figures and (ii) 4 decimal places.

$$37.81 \div 1.7 = \frac{37.81}{1.7} \ .$$

The denominator is changed into an integer by multiplying by 10. The numerator is also multiplied by 10 to keep the fraction the same. Thus

$$37.81 \div 1.7 = \frac{37.81 \times 10}{1.7 \ \times 10} = \frac{378.1}{17}$$

The long division is similar to the long division of integers and the first four steps are as shown:

```
        22.24117 . . . . .
 17) 378.10000
      34
      38
      34
      41
      34
      70
      68
      20
```

With reference to para 17,
(i) **37.81 ÷ 1.7 = 22.24, correct to 4 significant figures;** and
(ii) **37.81 ÷ 1.7 = 22.2412, correct to 4 decimal places.**

**Problem 38** Convert (a) 0.4375 to a proper fraction and (b) 4.285 to a mixed number.

(a) 0.4375 can be written as $\dfrac{0.437\ 5 \times 10\ 000}{10\ 000}$ without changing its value,
i.e. $0.437\ 5 = \dfrac{4\ 375}{10\ 000}$ .
By cancelling $\dfrac{4\ 375}{10\ 000} = \dfrac{875}{2\ 000} = \dfrac{175}{400} = \dfrac{35}{80} = \dfrac{7}{16}$
i.e. $\mathbf{0.437\ 5 = \dfrac{7}{16}}$ .

(b) Similarly, $4.285 = 4\dfrac{285}{1\ 000} = 4\dfrac{57}{200}$

14

(a) To convert a proper fraction to a decimal fraction, the numerator is divided by the denominator. Division by 16 can be done by the long division method, or, more simply, by dividing by 2 and then 8.

$$2\ \overline{\smash)9^{1}.00}^{\ 4.50}$$
$$8\ \overline{\smash)4.5^{5}0^{2}0^{4}0}^{\ 0.5\ 6\ 2\ 5}$$
Thus, $\frac{9}{16} = \mathbf{0.5625}$

(b) For mixed numbers, it is only necessary to convert the proper fraction part of the mixed number to a decimal fraction. Thus, dealing with the $\frac{7}{8}$ part gives:

$$8\ \overline{\smash)7.0^{6}0^{4}0}^{\ 0.8\ 7\ 5}$$
i.e. $\frac{7}{8} = 0.875$
Thus $5\frac{7}{8} = \mathbf{5.875}$

*Further problems on decimals may be found in section C, problems 54 to 68, page 18.*

## (g) PERCENTAGES

A decimal fraction is converted to a percentage by multiplying by 100. Thus,

(a) 1.875 corresponds to 1.875 × 100%, i.e., **187.5%**.
(b) 0.012 5 corresponds to 0.012 5 × 100%, i.e. **1.25%**.

To convert fractions to percentages, they are (i) converted to decimal fractions and (ii) multiplied by 100.

(a) By division, $\frac{5}{16} = 0.312\ 5$, hence $\frac{5}{16}$ corresponds to 0.312 5 × 100%, i.e., **31.25%**.
(b) Similarly, $1\frac{2}{5} = 1.4$ when expressed as a decimal fraction
Hence $1\frac{2}{5} = 1.4 × 100\% = 140\%$

$12\frac{1}{2}\%$ of £378 means $\dfrac{12\frac{1}{2}}{100} × 378$, since percent means 'per hundred'. Hence

$12\frac{1}{2}\%$ of £378 $= \dfrac{\overset{1}{\cancel{12\frac{1}{2}}}}{\underset{8}{\cancel{100}}} × 378 = \dfrac{378}{8} = \mathbf{£47.25}$

15

*Problem 43* Express 25 minutes as a percentage of 2 hours, correct to the nearest 1%

Working in minute units 2 hours = 120 minutes.
Hence 25 minutes is $\frac{25}{120}$ ths of 2 hours

By cancelling, $\frac{25}{120} = \frac{5}{24}$,

Expressing $\frac{5}{24}$ as a decimal fraction, gives $0.208\dot{3}$

Multiplying by 100 to convert the decimal fraction to a percentage gives:

$0.208\,\dot{3} \times 100 = 20.8\dot{3}\%$.

Thus **25 minutes is 21% of 2 hours**, correct to the nearest 1%.

*Problem 44* A German silver alloy consists of 60% copper, 25% zinc and 15% nickel. Determine the masses of the copper, zinc and nickel in a 3.74 kilogram block of the alloy.

By direct proportion:   100% corresponds to 3.74 kg

1% corresponds to $\frac{3.74}{100} = 0.0374$ kg

60% corresponds to 60 × 0.0374 = 2.244 kg

25% corresponds to 25 × 0.0374 = 0.935 kg

and   15% corresponds to 15 × 0.0374 = 0.561 kg.

Thus, the masses of the copper, zinc and nickel are **2.244 kg, 0.935 kg and 0.561 kg** respectively.
[Check: 2.244+0.935+0.561 = 3.74]

*Further problems on percentages may be found in section C following, problems 69 to 75, page 19.*

## C. FURTHER PROBLEMS ON ARITHMETIC

*Arithmetic operations.*
In *Problems 1 to 21*, determine the values of the expressions given.

| | | |
|---|---|---|
| 1 | 67−82+34 | [19] |
| 2 | 124−273+481−398 | [−66] |
| 3 | 927−114+182−183−247 | [565] |
| 4 | 2 417−487+2 424−1 778−4 712 | [−2 136] |
| 5 | −38 419−2 177+2 440−799+2 834 | [−36 121] |
| 6 | 2 715−18 250+11 471−1 509+113 274 | [107 701] |
| 7 | 73−57 | [16] |
| 8 | 813−(−674) | [1 487] |

16

| | | |
|---|---|---|
| 9 | 647−872 | [−225] |
| 10 | 3 151−(−2 763) | [5 914] |
| 11 | 4 872−4 683 | [189] |
| 12 | −23 148−47 724 | [−70 872] |
| 13 | 38 441−53 774 | [−15 333] |
| 14 | (a) 261 × 7; (b) 462 × 9 | [(a) 1 827; (b) 4 158] |
| 15 | (a) 783 × 11; (b) 73 × 24 | [(a) 8 613; (b) 1 752] |
| 16 | (a) 27 × 38; (b) 77 × 29 | [(a) 1 026; (b) 2 233] |
| 17 | (a) 448 × 23; (b) 143 × (−31) | [(a) 10 304; (b) −4 433] |
| 18 | (a) 288 ÷ 6; (b) 979 ÷ 11 | [(a) 48; (b) 89] |
| 19 | (a) 1 813/7; (b) 896/16 | [(a) 259; (b) 56] |
| 20 | (a) 21 432/47; (b) 15 904 ÷ 56 | [(a) 456; (b) 284] |
| 21 | (a) 88 738/187; (b) 46 857 ÷ 79 | [(a) $474\frac{100}{187}$ (b) $593\frac{10}{79}$] |

*Highest common factors and lowest common multiples*

In *Problems 22 to 27* find (a) the HCF and (b) the LCM of the numbers given.

| | | |
|---|---|---|
| 22 | 6, 10, 14 | [(a) 2; (b) 210] |
| 23 | 12, 30, 45 | [(a) 3; (b) 180] |
| 24 | 10, 15, 70, 105 | [(a) 5; (b) 210] |
| 25 | 90, 105, 300 | [(a) 15; (b) 6 300] |
| 26 | 210, 196, 910, 462 | [(a) 14; (b) 420 420] |
| 27 | 196, 350, 770 | [(a) 14; (b) 53 900] |

*Order of precedence and brackets*

Simplify the expressions given in *Problems 28 to 34.*

| | | |
|---|---|---|
| 28 | 14+3 × 15 | [59] |
| 29 | 17−12 ÷ 4 | [14] |
| 30 | 86+24 ÷(14−2) | [88] |
| 31 | 7(23−18) ÷ (12−5) | [5] |
| 32 | 63−28(14 ÷ 2) + 26 | [−107] |
| 33 | 112/16−119 ÷ 17+(3 × 19) | [57] |
| 34 | (50−14)/3+7(16−7)−7 | [68] |

*Fractions*

Evaluate the expressions given in *Problems 35 to 47*

| | | |
|---|---|---|
| 35 | (a) $\frac{1}{2} + \frac{2}{5}$   (b) $\frac{7}{16} - \frac{1}{4}$ | [(a) $\frac{9}{10}$; (b) $\frac{3}{16}$] |
| 36 | (a) $\frac{2}{7} + \frac{3}{11}$  (b) $\frac{2}{9} - \frac{1}{7} + \frac{2}{3}$ | [(a) $\frac{43}{77}$; (b) $\frac{47}{63}$] |
| 37 | (a) $5\frac{3}{13} + 3\frac{3}{4}$   (b) $4\frac{5}{8} - 3\frac{2}{5}$ | [(a) $8\frac{51}{52}$; (b) $1\frac{9}{40}$] |
| 38 | (a) $10\frac{3}{7} - 8\frac{2}{3}$   (b) $3\frac{1}{4} - 4\frac{4}{5} + 1\frac{5}{6}$ | [(a) $1\frac{16}{21}$; (b) $\frac{17}{60}$] |
| 39 | (a) $\frac{3}{4} \times \frac{5}{9}$   (b) $\frac{17}{35} \times \frac{15}{119}$ | [(a) $\frac{5}{12}$; (b) $\frac{3}{49}$] |
| 40 | (a) $\frac{3}{5} \times \frac{7}{9} \times 1\frac{2}{7}$   (b) $\frac{13}{17} \times 4\frac{7}{11} \times 3\frac{4}{39}$ | [(a) $\frac{3}{5}$; (b) 11] |

41 (a) $\frac{1}{4} \times \frac{3}{11} \times 1\frac{5}{39}$   (b) $\frac{3}{4} \div 1\frac{4}{5}$                $[(a)\ \frac{1}{13}; (b)\ \frac{5}{12}]$

42 (a) $\frac{3}{8} \div \frac{45}{64}$   (b) $1\frac{1}{3} \div 2\frac{5}{9}$                $[(a)\ \frac{8}{15}; (b)\ \frac{12}{23}]$

43 $\frac{1}{3} - \frac{3}{4} \times \frac{16}{27}$                $[-\frac{1}{9}]$

44 $\frac{1}{2} + \frac{3}{5} \div \frac{9}{15} - \frac{1}{3}$                $[1\frac{1}{6}]$

45 $\frac{7}{15}$ of $\left(15 \times \frac{5}{7}\right) + \left(\frac{3}{4} \div \frac{15}{16}\right)$                $[5\frac{4}{5}]$

46 $\frac{1}{4} \times \frac{2}{3} - \frac{1}{3} \div \frac{3}{5} + \frac{2}{7}$                $[-\frac{13}{126}]$

47 $\left(\frac{2}{3} \times 1\frac{1}{4}\right) \div \left(\frac{2}{3} + \frac{1}{4}\right) + 1\frac{3}{5}$                $[2\frac{28}{55}]$

*Ratio and proportion*

48 Divide 312 in the ratio of 7 to 17.                [91 to 221]

49 Divide 621 in the ratio of 3 to 7 to 13.                [81 to 189 to 351]

50 £4.94 is to be divided between two people in the ratio of 9 to 17.
Determine how much each person will receive.                [£1.71 and £3.23]

51 When mixing a quantity of paints, dyes of four different colours are used in the ratio of 7 : 3 : 19 : 5. If the mass of the first dye used is $3\frac{1}{2}$ g, determine the total mass of the dyes used.                [17 g]

52 It takes 21 hours for 12 men to resurface a stretch of road. Find how many men it takes to resurface a similar stretch of road in 50 hours 24 minutes, assuming the work rate remains constant.                [5]

53 It takes 3 hours 15 minutes to fly from city A to city B at a constant speed. Find how long the journey takes if (a) the speed if $1\frac{1}{2}$ times that of the original speed and (b) if the speed is three-quarters of the original speed.
                [(a) 2 h 10 min; (b) 4 h 20 min]

*Decimals*

In *Problems 54 to 60*, determine the values of the expressions given.

54 $23.6 + 14.71 - 18.9 - 7.421$                [11.989]

55 $73.84 - 113.247 + 8.21 - 0.068$                [−31.265]

56 $5.73 \times 4.2$                [24.066]

57 $3.8 \times 4.1 \times 0.7$                [10.906]

58 $374.1 \times 0.006$                [2.244 6]

59 $421.8 \div 17$, (a) correct to 4 significant figures and (b) correct to 3 decimal places.
                [(a) 24.81; (b) 24.812]

60 0.014 7/2.3, (a) correct to 5 decimal places and (b) correct to 2 significant figures.
                [(a) 0.006 39; (b) 0.006 4]

61 Convert to proper fractions:
(a) 0.65; (b) 0.84; (c) 0.012 5; (d) 0.282 and (e) 0.024.

$$\left[(a)\ \frac{13}{20}; (b)\ \frac{21}{25}; (c)\ \frac{1}{80}; (d)\ \frac{141}{500}; (e)\ \frac{3}{125}\right]$$

62 Convert to mixed numbers:
(a) 1.82; (b) 4.275; (c) 14.125; (d) 15.35 and (e) 16.212 5.

$$[(a)\ 1\frac{41}{50}; (b)\ 4\frac{11}{40}; (c)\ 14\frac{1}{8}; (d)\ 15\frac{7}{20}; (e)\ 16\frac{17}{80}]$$

In problems 63 to 68, express as decimal fractions to the accuracy stated.

63 $\frac{4}{9}$, correct to 5 significant figures.                    [0.444 44]

64 $\frac{17}{27}$, correct to 5 decimal places.                    [0.629 63]

65 $1\frac{9}{16}$, correct to 4 significant figures                    [1.563]

66 $53\frac{5}{11}$, correct to 3 decimal places.                    [53.455]

67 $13\frac{31}{37}$, correct to 2 decimal places.                    [13.84]

68 $8\frac{9}{13}$, correct to 3 significant figures                    [8.69]

*Percentages*

69 Convert to percentages: (a) 0.057; (b) 0.374 and (c) 1.285.

[(a) 5.7%; (b) 37.4%; (c) 128.5%]

70 Express as percentages, correct to 3 significant figures:

(a) $\frac{7}{33}$; (b) $\frac{19}{24}$ and (c) $1\frac{11}{16}$.

[(a) 21.2%; (b) 79.2%; (c) 169%]

71 Calculate correct to 4 significant figures:
   (a) 18% of 2 758 tonnes;
   (b) 47% of 18.42 grams; and
   (c) 147% of 14.1 seconds.

[(a) 496.4 t; (b) 8.657 g; (c) 20.73 s]

72 Express:
   (a) 140 kg as a percentage of 1 t;
   (b) 47 s as a percentage of 5 min; and
   (c) 13.4 cm as a percentage of 2.5 m.

[(a) 14%; (b) 15.6%; (c) 5.36%]

73 A block of monel alloy consists of 70% nickel and 30% copper. If it contains 88.2 g of nickel, determine the mass of copper in the block.

[37.8 g]

74 Two kilograms of a compound contains 30% of element A, 45% of element B and 25% of element C. Determine the masses of the three elements present.

[A 0.6 kg; B 0.9 kg; C 0.5 kg]

75 A concrete mixture contains seven parts by volume of ballast, four parts by volume of sand and two parts by volume of cement. Determine the percentage of each of these three constituents correct to the nearest 1% and the mass of cement in a two tonne dry mix, correct to 1 significant figure.

[54%; 31%; 15%; 0.3 t]

# 2 Indices, standard and binary forms

## A. MAIN POINTS CONCERNING INDICES, STANDARD AND BINARY FORMS

1   The lowest factors of 2000 are $2 \times 2 \times 2 \times 2 \times 5 \times 5 \times 5$. These factors are written as $2^4 \times 5^3$, where 2 and 5 are called **bases** and the numbers 4 and 3 are called **indices**. When an index is an integer it is called a **power**. Thus, $2^4$ is called 'two to the power of four', and has a base of 2 and an index of 4. Similarly, $5^3$ is called 'five to a power of 3' and has a base of 5 and an index of 3. Special names may be used when the indices are 2 and 3, these being called 'squared' and 'cubed' respectively. Thus $7^2$ is called 'seven squared' and $9^3$ is called 'nine cubed'. When no index is shown, the power is 1, i.e. $2^1$ means 2.

2   The **reciprocal** of a number is when the index is $-1$ and its value is given by 1 divided by the base. Thus the reciprocal of 2 is $2^{-1}$ and its value is $1/2$ or 0.5. Similarly, the reciprocal of 5 is $5^{-1}$ which means $1/5$ or 0.2.

3   The **square root** of a number is when the index is $1/2$, and the square root of 2 is written as $2^{\frac{1}{2}}$ or $\sqrt{2}$. The value of a square root is the value of the base which when multiplied by itself gives the number. Since $3 \times 3 = 9$, then $\sqrt{9} = 3$. However, $(-3) \times (-3) = 9$, so $\sqrt{9} = -3$. There are always two answers when finding the square root of a number and this is shown by putting both a $+$ and a $-$ sign in front of the answer to a square root problem. Thus $\sqrt{9} = \pm 3$ and $4^{\frac{1}{2}} = \sqrt{4} = \pm 2$, and so on.

4   When simplifying calculations involving indices, certain basic rules or laws can be applied, called the **laws of indices**. These are given below.

(a) When multiplying two or more numbers having the same base, the indices are added. Thus $3^2 \times 3^4 = 3^{2+4} = 3^6$.

(b) When a number is divided by a number having the same base, the indices are subtracted. Thus

$$\frac{3^5}{3^2} = 3^{5-2} = 3^3.$$

(c) When a number which is raised to a power is raised to a further power, the indices are multiplied. Thus $(3^5)^2 = 3^{5 \times 2} = 3^{10}$.

(d) When a number has an index of 0, its value is 1. Thus $3^0 = 1$.

(e) A number raised to a negative power is the reciprocal of that number raised to a positive power. Thus $3^{-4} = \frac{1}{3^4}$. Similarly, $\frac{1}{2^{-3}} = 2^3$, (see *Problem 13*)

(f) when a number is raised to a fractional power the denominator of the fraction is the root of the number and the numerator is the power. Thus

$$8^{\frac{2}{3}} = \sqrt[3]{8^2} = (2)^2 = 4$$

and

$$25^{\frac{1}{2}} = \sqrt{25^1} = \pm 5$$

5    A number written with one digit to the left of the decimal point and multiplied by 10 raised to some power is said to be written in **standard form**. Thus:
5 837 is written as $5.837 \times 10^3$ in standard form, and
0.0415 is written as $4.15 \times 10^{-2}$ in standard form.
When a number is written in standard form, the first factor is called the **mantissa** and the second factor is called the **exponent**. Thus the number $5.8 \times 10^3$ has a mantissa of 5.8 and an exponent of $10^3$.

6    (i) Numbers having the same exponent can be added or subtracted in standard form by adding or subtracting the mantissae and keeping the exponent the same. Thus:
$2.3 \times 10^4 + 3.7 \times 10^4 = (2.3 + 3.7) \times 10^4 = 6.0 \times 10^4$, and
$5.9 \times 10^{-2} - 4.6 \times 10^{-2} = (5.9 - 4.6) \times 10^{-2} = 1.3 \times 10^{-2}$
When numbers have different exponents, one way of adding or subtracting the numbers is to express one of the numbers in non-standard form, so that both numbers have the same exponent. Thus:

$$2.3 \times 10^4 + 3.7 \times 10^3 = 2.3 \times 10^4 + 0.37 \times 10^4$$
$$= (2.3 + 0.37) \times 10^4 = 2.67 \times 10^4$$

Alternatively, $2.3 \times 10^4 + 3.7 \times 10^3 = 23\,000 + 3\,700 = 26\,700 = 2.67 \times 10^4$

(ii) The laws of indices are used when multiplying or dividing numbers given in standard form. For example,

$$(2.5 \times 10^3) \times (5 \times 10^2) = (2.5 \times 5) \times (10^{3+2}) = 12.5 \times 10^5 \text{ or } 1.25 \times 10^6.$$

Similarly,

$$\frac{6 \times 10^4}{1.5 \times 10^2} = \frac{6}{1.5} \times (10^{4-2}) = 4 \times 10^2$$

7    The ordinary system of numbers in everyday use is the **denary** or **decimal system of numbers**, using the digits 0 to 9. It has ten different digits, (0, 1, 2, 3, 4, 5, 6, 7, 8, 9), and is said to have a radix of 10. The **binary system of numbers** has a radix of 2 and uses only the digits 0 and 1. The binary number 1101 means

$$1 \times 2^3 + 1 \times 2^2 + 0 \times 2^1 + 1 \times 2^0$$
$$= 8 \quad + \quad 4 \quad + \quad 0 \quad + \quad 1 \quad = 13$$

Hence $1101_2 = 13_{10}$, the suffix 2 and suffix 10 being used to avoid confusion between binary and denary numbers. Similarly

$$110101_2 = 1 \times 2^5 + 1 \times 2^4 + 0 \times 2^3 + 1 \times 2^2 + 0 \times 2^1 + 1 \times 2^0$$
$$= 32 \quad + 16 \quad + 0 \quad + 4 \quad + 0 \quad + 1$$
$$= 53_{10}$$

8    A denary number can be converted into a corresponding binary number by repeatedly dividing by 2 and recording the remainder at each stage. The corresponding binary number is given by the remainder column read from the bottom upwards, (see *Problems 22 and 23*).

9    When adding two binary numbers, if the sum of a column is 2, then this is shown as having a sum of 0 and 1 is carried to the next column on the left. If the sum of a column is 3, (due to 1 being carried from the previous column and the two binary digits (bits) being added), then this is shown as having a sum of 1, and 1 is

carried to the next column on the left. Thus, adding $101_2$ and $111_2$ gives:

```
      0101
      0111
sum   1100  ,   i.e. 101₂ + 111₂ = 1100₂
carry 111
```

sum $\overline{1100}$ , i.e. $101_2 + 111_2 = 1100_2$

10 Because a binary system of numbers has only two digits, 0 and 1, it can be readily related to other systems which have only two states. For example, ON/OFF, TRUE/FALSE, YES/NO, PRESENT/ABSENT, DEFECTIVE/NON-DEFECTIVE, and so on. A binary system is used as a basis for coding computers, calculators and microprocessors and is also widely used in telecommunications engineering.

## B. WORKED PROBLEMS ON INDICES, STANDARD AND BINARY FORMS

### (a) INDICES

*Problem 1* Evaluate: (a) $5^2 \times 5^3$, (b) $3^2 \times 3^4 \times 3$ and (c) $2 \times 2^2 \times 2^5$

From para 4(a):

(a) $5^2 \times 5^3 = 5^{(2+3)} = 5^5 = 5 \times 5 \times 5 \times 5 \times 5 = \mathbf{3125}$
(b) $3^2 \times 3^4 \times 3 = 3^{(2+4+1)} = 3^7 = 3 \times 3 \times \ldots \text{ to 7 terms} = \mathbf{2187}$
(c) $2 \times 2^2 \times 2^5 = 2^{(1+2+5)} = 2^8 = \mathbf{256}$

*Problem 2* Find the value of: (a) $\dfrac{7^5}{7^3}$ and (b) $\dfrac{5^7}{5^4}$

From para. 4(b):

(a) $\dfrac{7^5}{7^3} = 7^{(5-3)} = 7^2 = \mathbf{49}$

(b) $\dfrac{5^7}{5^4} = 5^{(7-4)} = 5^3 = \mathbf{125}$

*Problem 3* Evaluate: (a) $5^2 \times 5^3 \div 5^4$ and (b) $(3 \times 3^5) \div (3^2 \times 3^3)$

From para 4, (a) and (b):

$$\begin{aligned}
\text{(a) } 5^2 \times 5^3 \div 5^4 \quad &= \frac{5^2 \times 5^3}{5^4} = \frac{5^{(2+3)}}{5^4} \\
&= \frac{5^5}{5^4} = 5^{(5-4)} \\
&= 5^1 = \mathbf{5}
\end{aligned}$$

22

(b) $(3 \times 3^5) \div (3^2 \times 3^3) = \dfrac{3 \times 3^5}{3^2 \times 3^3} = \dfrac{3^{(1+5)}}{3^{(2+3)}}$

$= \dfrac{3^6}{3^5} = 3^{6-5}$

$= 3^1 = 3$

**Problem 4** Simplify: (a) $(2^3)^4$ and (b) $(3^2)^5$, expressing the answers in index form.

From para 4(c):

(a) $(2^3)^4 = 2^{3 \times 4} = \mathbf{2^{12}}$
(b) $(3^2)^5 = 3^{2 \times 5} = \mathbf{3^{10}}$

**Problem 5** Evaluate: $\dfrac{(10^2)^3}{10^4 \times 10^2}$

From para 4:

$\dfrac{(10^2)^3}{10^4 \times 10^2} = \dfrac{10^{(2 \times 3)}}{10^{(4+2)}} = \dfrac{10^6}{10^6} = 10^{6-6} = 10^0 = \mathbf{1}$

**Problem 6** Find the value of (a) $\dfrac{2^3 \times 2^4}{2^7 \times 2^5}$ and (b) $\dfrac{(3^2)^3}{3 \times 3^9}$

From para 4:

(a) $\dfrac{2^3 \times 2^4}{2^7 \times 2^5} = \dfrac{2^{(3+4)}}{2^{(7+5)}} = \dfrac{2^7}{2^{12}}$

$= 2^{7-12} = 2^{-5} = \dfrac{1}{2^5} = \mathbf{\dfrac{1}{32}}$

(b) $\dfrac{(3^2)^3}{3 \times 3^9} = \dfrac{3^{2 \times 3}}{3^{1+9}} = \dfrac{3^6}{3^{10}}$

$= 3^{6-10} = 3^{-4} = \dfrac{1}{3^4} = \mathbf{\dfrac{1}{81}}$

**Problem 7** Evaluate (a) $4^{\frac{1}{2}}$; (b) $16^{\frac{3}{4}}$; (c) $27^{\frac{2}{3}}$; (d) $9^{-\frac{1}{2}}$

(a) $4^{\frac{1}{2}} = \sqrt{4} = \pm 2$ 　　　　(b) $16^{\frac{3}{4}} = \sqrt[4]{16^3} = (2)^3 = \mathbf{8}$

(Note that it does not matter whether the 4th root of 16 is found first or whether 16 cubed is found first—the same answer will result.)

(c) $27^{\frac{2}{3}} = \sqrt[3]{27^2} = (3)^2 = \mathbf{9}$ 　　(d) $9^{-\frac{1}{2}} = \dfrac{1}{9^{\frac{1}{2}}} = \dfrac{1}{\sqrt{9}} = \dfrac{1}{\pm 3} = \pm \dfrac{1}{3}$

23

The laws of indices only apply to terms **having the same base**. Grouping terms having the same base, and then applying the laws of indices to each of the groups independently gives:

$$\frac{3^3 \times 5^7}{5^3 \times 3^4} = \frac{3^3}{3^4} \times \frac{5^7}{5^3}$$

$$= 3^{(3-4)} \times 5^{(7-3)} = 3^{-1} \times 5^4$$

$$= \frac{5^4}{3^1} = \frac{625}{3} = 208\frac{1}{3}$$

$$\frac{2^3 \times 3^5 \times (7^2)^2}{7^4 \times 2^4 \times 3^3} = 2^{3-4} \times 3^{5-3} \times 7^{2 \times 2 - 4}$$

$$= 2^{-1} \times 3^2 \times 7^0 = \frac{1}{2} \times 3^2 \times 1$$

$$= \frac{9}{2} = 4\frac{1}{2}$$

$$4^{1.5} = 4^{\frac{3}{2}} = \sqrt{4^3} = 2^3 = 8, \quad 8^{\frac{1}{3}} = \sqrt[3]{8} = 2, \quad 2^2 = 4,$$

$$32^{-\frac{2}{5}} = \frac{1}{32^{\frac{2}{5}}} = \frac{1}{\sqrt[5]{32^2}} = \frac{1}{2^2} = \frac{1}{4}$$

Hence $\dfrac{4^{1.5} \times 8^{\frac{1}{3}}}{2^2 \times 32^{-\frac{2}{5}}} = \dfrac{8 \times 2}{4 \times \frac{1}{4}} = \dfrac{16}{1} = \mathbf{16}$

Alternatively, $\dfrac{4^{1.5} \times 8^{\frac{1}{3}}}{2^2 \times 32^{-\frac{2}{5}}} = \dfrac{[(2)^2]^{\frac{3}{2}} \times (2^3)^{\frac{1}{3}}}{2^2 \times (2^5)^{-\frac{2}{5}}} = \dfrac{2^3 \times 2^1}{2^2 \times 2^{-2}} = 2^{3+1-2-(-2)}$

$$= 2^4 = 16$$

**Problem 11** Evaluate: $\dfrac{3^2 \times 5^5 + 3^3 \times 5^3}{3^4 \times 5^4}$

Dividing each term by the HCF of the three terms, i.e., $3^2 \times 5^3$, gives:

$$\frac{3^2 \times 5^5 + 3^3 \times 5^3}{3^4 \times 5^4} = \frac{\dfrac{3^2 \times 5^5}{3^2 \times 5^3} + \dfrac{3^3 \times 5^3}{3^2 \times 5^3}}{\dfrac{3^4 \times 5^4}{3^2 \times 5^3}}$$

$$= \frac{3^{(2-2)} \times 5^{(5-3)} + 3^{(3-2)} \times 5^{(3-3)}}{3^{(4-2)} \times 5^{(4-3)}}$$

$$= \frac{3^0 \times 5^2 + 3^1 \times 5^0}{3^2 \times 5^1}$$

$$= \frac{1 \times 25 + 3 \times 1}{9 \times 5} = \frac{28}{45}$$

**Problem 12** Find the value of $\dfrac{3^2 \times 5^5}{3^4 \times 5^4 + 3^3 \times 5^3}$

To simplify the arithmetic, each term is divided by the HCF of all the terms, i.e. $3^2 \times 5^3$. Thus

$$\frac{3^2 \times 5^5}{3^4 \times 5^4 + 3^3 \times 5^3} = \frac{\dfrac{3^2 \times 5^5}{3^2 \times 5^3}}{\dfrac{3^4 \times 5^4}{3^2 \times 5^3} + \dfrac{3^3 \times 5^3}{3^2 \times 5^3}} = \frac{3^{(2-2)} \times 5^{(5-3)}}{3^{(4-2)} \times 5^{(4-3)} + 3^{(3-2)} \times 5^{(3-3)}}$$

$$= \frac{3^0 \times 5^2}{3^2 \times 5^1 + 3^1 \times 5^0} = \frac{1 \times 5^2}{3^2 \times 5 + 3 \times 1}$$

$$= \frac{25}{45+3} \qquad\qquad = \frac{25}{48}$$

**Problem 13** Simplify $\dfrac{7^{-3} \times 3^4}{3^{-2} \times 7^5 \times 5^{-2}}$, expressing the answer in index form with positive indices.

Since $7^{-3} = \dfrac{1}{7^3}$, $\dfrac{1}{3^{-2}} = 3^2$ and $\dfrac{1}{5^{-2}} = 5^2$, then

$$\frac{7^{-3} \times 3^4}{3^{-2} \times 7^5 \times 5^{-2}} = \frac{3^4 \times 3^2 \times 5^2}{7^3 \times 7^5} = \frac{3^{(4+2)} \times 5^2}{7^{(3+5)}} = \frac{3^6 \times 5^2}{7^8}$$

25

Expressing the numbers in terms of their lowest prime numbers gives:

$$\frac{16^2 \times 9^{-2}}{4 \times 3^3 - 2^{-3} \times 8^2} = \frac{(2^4)^2 \times (3^2)^{-2}}{2^2 \times 3^3 - 2^{-3} \times (2^3)^2}$$

$$= \frac{2^8 \times 3^{-4}}{2^2 \times 3^3 - 2^{-3} \times 2^6} = \frac{2^8 \times 3^{-4}}{2^2 \times 3^3 - 2^3}$$

Dividing each term by the HCF, (i.e. $2^2$), gives: $\dfrac{2^6 \times 3^{-4}}{3^3 - 2} = \dfrac{2^6}{3^4(3^3 - 2)}$

A fraction raised to a power means that both the numerator and the denominator of the fraction are raised to that power, i.e.

$$\left(\frac{4}{3}\right)^3 = \frac{4^3}{3^3}$$

A fraction raised to a negative power has the same value as the inverse of the fraction raised to a positive power.

Thus, $\left(\dfrac{3}{5}\right)^{-2} = \dfrac{1}{\left(\dfrac{3}{5}\right)^2} = \dfrac{1}{\dfrac{3^2}{5^2}} = 1 \times \dfrac{5^2}{3^2} = \dfrac{5^2}{3^2} = \left(\dfrac{5}{3}\right)^2$.

Similarly, $\left(\dfrac{2}{5}\right)^{-3} = \left(\dfrac{5}{2}\right)^3$.

Thus, $\dfrac{\left(\dfrac{4}{3}\right)^3 \times \left(\dfrac{3}{5}\right)^{-2}}{\left(\dfrac{2}{5}\right)^{-3}} = \dfrac{\dfrac{4^3}{3^3} \times \dfrac{5^2}{3^2}}{\dfrac{5^3}{2^3}} = \dfrac{4^3}{3^3} \times \dfrac{5^2}{3^2} \times \dfrac{2^3}{5^3}$

$$= \frac{(2^2)^3 \times 2^3}{3^{(3+2)} \times 5^{(3-2)}} = \frac{2^9}{3^5 \times 5}$$

*Further problems on indices may be found in section C, problems 1 to 22, page 30.*

**Problem 16** Express in standard form: (a) 38.71; (b) 3 746 and (c) 0.0124

For a number to be in standard form, it is expressed with only one digit to the left of the decimal point. Thus:
(a) 38.71 must be divided by 10 to achieve one digit to the left of the decimal point and it must also be multiplied by 10 to maintain the equality, i.e.

$$38.71 = \frac{38.71}{10} \times 10 = \mathbf{3.871 \times 10} \text{ in standard form}$$

(b) $3\ 746 = \frac{3\ 746}{1\ 000} \times 1\ 000 = \mathbf{3.746 \times 10^3}$ in standard form

(c) $0.0124 = 0.0124 \times \frac{100}{100} = \frac{1.24}{100} = \mathbf{1.24 \times 10^{-2}}$ in standard form

**Problem 17** Express the following numbers, which are in standard form, as decimal numbers. (a) $1.725 \times 10^{-2}$, (b) $5.491 \times 10^4$ and (c) $9.84 \times 10^0$.

(a) $1.725 \times 10^{-2} = \frac{1.725}{100} = \mathbf{0.017\ 25}$

(b) $5.491 \times 10^4 = 5.491 \times 10\ 000 = \mathbf{54\ 910}$

(c) $9.84 \times 10^0 = 9.84 \times 1 = \mathbf{9.84}$ (since $10^0 = 1$)

**Problem 18** Express in standard form, correct to 3 significant figures.
(a) $\frac{3}{8}$, (b) $19\frac{2}{3}$ and (c) $741\frac{9}{16}$.

(a) $\frac{3}{8} = 0.375$, and expressing it in standard form gives:
$$0.375 = \mathbf{3.75 \times 10^{-1}}$$
(b) $19\frac{2}{3} = 19.\dot{6} = \mathbf{1.97 \times 10}$ in standard form, correct to 3 significant figures.
(c) $741\frac{9}{16} = 741.5\ 625 = \mathbf{7.42 \times 10^2}$ in standard form, correct to 3 significant figures.

**Problem 19** Express the following numbers, given in standard form, as fractions or mixed numbers: (a) $2.5 \times 10^{-1}$, (b) $6.25 \times 10^{-2}$ and (c) $1.354 \times 10^2$.

(a) $2.5 \times 10^{-1} = \frac{2.5}{10} = \frac{25}{100} = \mathbf{\frac{1}{4}}$

(b) $6.25 \times 10^{-2} = \frac{6.25}{100} = \frac{625}{10\ 000} = \mathbf{\frac{1}{16}}$

(c) $1.354 \times 10^2 = 135.4 = 135\frac{4}{10} = \mathbf{135\frac{2}{5}}$

**Problem 20** Find the value of (a) $7.9 \times 10^{-2} - 5.4 \times 10^{-2}$, (b)
(b) $8.3 \times 10^3 + 5.415 \times 10^3$ and (c) $9.293 \times 10^2 + 1.3 \times 10^3$, expressing the
answers in standard form.

Numbers having the same exponent can be added or subtracted by adding or subtracting the mantissae and keeping the exponent the same. Thus:

(a) $7.9 \times 10^{-2} - 5.4 \times 10^{-2}$ $= (7.9 - 5.4) \times 10^{-2}$
$= \mathbf{2.5 \times 10^{-2}}$

(b) $8.3 \times 10^3 + 5.415 \times 10^3$ $= (8.3 + 5.415) \times 10^3$
$= 13.715 \times 10^3$
$= \mathbf{1.3715 \times 10^4}$ in standard form.

(c) Since only numbers having the same exponents can be added by straight addition of the mantissae, the numbers are converted to this form before adding. Thus:

$9.293 \times 10^2 + 1.3 \times 10^3$ $= 9.293 \times 10^2 + 13 \times 10^2$
$= (9.293 + 13) \times 10^2$
$= 22.293 \times 10^2$
$= \mathbf{2.229\ 3 \times 10^3}$ in standard form.

Alternatively, the numbers can be expressed as decimal fractions, giving

$9.293 \times 10^2 + 1.3 \times 10^3$ $= 929.3 + 1\ 300$
$= 2\ 229.3$
$= \mathbf{2.229\ 3 \times 10^3}$ in standard form,

as obtained previously. This method is often the 'safest' way of doing this type of problem.

**Problem 21** Evaluate (a) $(3.75 \times 10^3)(6 \times 10^4)$
and (b) $\dfrac{3.5 \times 10^5}{7 \times 10^2}$, expressing answers in standard form.

(a) $(3.75 \times 10^3)(6 \times 10^4) = (3.75 \times 6)(10^{3+4}) = 22.50 \times 10^7 = \mathbf{2.25 \times 10^8}$

(b) $\dfrac{3.5 \times 10^5}{7 \times 10^2} = \dfrac{3.5}{7} \times 10^{5-2} = 0.5 \times 10^3 = \mathbf{5 \times 10^2}$.

*Further problems on standard form may be found in section C, problems 23 to 34, page 31.*

(c) BINARY FORM

**Problem 22** Express the denary number 14 as a binary number.

The remainder method is used, in which the denary number is repeatedly divided by 2, the remainder being noted at each stage.

28

| 2 ) 14 | |
|--------|---|
| 2 ) 7 | 0 |
| 2 ) 3 | 1 |
| 2 ) 1 | 1 |
| 0 | 1 |

Binary Number

The corresponding binary number is the value of the remainder column read from the bottom upwards.
Thus $14_{10} = 1110_2$

**Problem 23** Find the binary number corresponding to the denary number 49

Using the remainder method:

Remainder

| 2 ) 49 | |
|--------|---|
| 2 ) 24 | 1 |
| 2 ) 12 | 0 |
| 2 ) 6 | 0 |
| 2 ) 3 | 0 |
| 2 ) 1 | 1 |
| 0 | 1 |

Binary Number

Reading the binary number from the bottom of the remainder column upwards, gives:

$49_{10} = 110\,001_2$

**Problem 24** Express the binary number 1100 as a denary number

The denary number, say, 357, means $3 \times 10^2 + 5 \times 10^1 + 7 \times 10^0$. Similarly, the binary number 1100 means $1 \times 2^3 + 1 \times 2^2 + 0 \times 2^1 + 0 \times 2^0$
i.e. $1100_2 = 8 + 4 + 0 + 0 = 12_{10}$

**Problem 25** Convert the number $101\,011_2$ to a denary number

$$
\begin{aligned}
101\,011_2 &= 1 \times 2^5 + 0 \times 2^4 + 1 \times 2^3 + 0 \times 2^2 + 1 \times 2^1 + 1 \times 2^0 \\
&= 1 \times 32 + 0 \times 16 + 1 \times 8 + 0 \times 4 + 1 \times 2 + 1 \times 1 \\
&= 32 + 0 + 8 + 0 + 2 + 1 = 43_{10}
\end{aligned}
$$

**Problem 26** Add the binary numbers 111 and 110

The numbers are laid-out as for denary addition. The columns are added, starting from the right. The bits of the two numbers being added in each column and the carry bit, when there is one, are added. With reference to para 9:

|        | 0 | 1 | 1 | 1 |
|--------|---|---|---|---|
|        | 0 | 1 | 1 | 0 |
| Sum:   | 1 | 1 | 0 | 1 |
| Carry: | 1 | 1 | 0 | |
| Step:  | (4) | (3) | (2) | (1) |

Step (1) $1+0 = 1$, carry 0
Step (2) $1+1+0 = 0$, carry 1
Step (3) $1+1+1 = 1$, carry 1
Step (4) $0+0+1 = 1$

Hence, $111_2 + 110_2 = 1101_2$

*Problem 27* Find the value of $1011_2 + 1101_2$, expressing the answer in both binary and denary forms

Using the rules for bit addition given in para 9:

|        | 0 | 1 | 0 | 1 | 1 |
|--------|---|---|---|---|---|
|        | 0 | 1 | 1 | 0 | 1 |
| Sum:   | 1 | 1 | 0 | 0 | 0 |
| Carry: | 1 | 1 | 1 | 1 |   |
| Step:  | (5) | (4) | (3) | (2) | (1) |

Step (1):   $1+1 = 0$, carry 1
Step (2): $1+1+0 = 0$, carry 1
Step (3): $1+0+1 = 0$, carry 1
Step (4): $1+1+1 = 1$, carry 1
Step (5): $1+0+0 = 1$, carry 0

Thus,   $1011_2 + 1101_2 = 11000_2$

$11000_2 = 1 \times 2^4 + 1 \times 2^3 = 16 + 8 = 24_{10}$

*Further problems on binary form may be found in section C following, problems 35 to 44, page 32.*

## C. FURTHER PROBLEMS ON INDICES, STANDARD AND BINARY FORMS

*Indices*

In *Problems 1 to 14*, simplify the expressions given, expressing the answers in index form and with positive indices.

1   (a) $3^3 \times 3^4$    (b) $4^2 \times 4^3 \times 4^4$        [(a) $3^7$, (b) $4^9$]

2   (a) $2^3 \times 2 \times 2^2$    (b) $7^2 \times 7^4 \times 7 \times 7^3$      [(a) $2^6$, (b) $7^{10}$]

3   (a) $\dfrac{2^4}{2^3}$    (b) $\dfrac{3^7}{3^2}$        [(a) 2, (b) $3^5$]

4   (a) $5^6 \div 5^3$    (b) $7^{13}/7^{10}$        [(a) $5^3$, (b) $7^3$]

5   (a) $(7^2)^3$    (b) $(3^3)^2$        [(a) $7^6$, (b) $3^6$]

6   (a) $(15^3)^5$    (b) $(17^2)^4$        [(a) $15^{15}$, (b) $17^8$]

7   (a) $\dfrac{2^2 \times 2^3}{2^4}$    (b) $\dfrac{3^7 \times 3^4}{3^5}$        [(a) 2, (b) $3^6$]

8   (a) $\dfrac{5^7}{5^2 \times 5^3}$    (b) $\dfrac{13^5}{13 \times 13^2}$        [(a) $5^2$, (b) $13^2$]

9   (a) $\dfrac{(9 \times 3^2)^3}{(3 \times 27)^2}$    (b) $\dfrac{(16 \times 4)^2}{(2 \times 8)^3}$        [(a) $3^4$, (b) 1]

10   (a) $\dfrac{5^{-2}}{5^{-4}}$    (b) $\dfrac{3^2 \times 3^{-4}}{3^3}$        [(a) $5^2$, (b) $\dfrac{1}{3^5}$]

11   (a) $\dfrac{7^2 \times 7^{-3}}{7 \times 7^{-4}}$    (b) $\dfrac{2^3 \times 2^{-4} \times 2^5}{2 \times 2^{-2} \times 2^6}$        [(a) $7^2$, (b) $\dfrac{1}{2}$]

12   (a) $13 \times 13^{-2} \times 13^4 \times 13^{-3}$    (b) $\dfrac{5^{-7} \times 5^2}{5^{-8} \times 5^3}$        [(a) 1, (b) 1]

13 (a) $\dfrac{3^3 \times 5^2}{5^4 \times 3^4}$     (b) $\dfrac{7^{-2} \times 3^{-2}}{3^5 \times 7^4 \times 7^{-3}}$     $\left[(a)\dfrac{1}{3 \times 5^2} \text{ (b) } \dfrac{1}{7^3 \times 3^7}\right]$

14 (a) $\dfrac{4^2 \times 9^3}{8^3 \times 3^4}$     (b) $\dfrac{8^{-2} \times 5^2 \times 3^{-4}}{25^2 \times 2^4 \times 9^{-2}}$     $\left[(a) \dfrac{3^2}{2^5} \text{ (b) } \dfrac{1}{2^{10} \times 5^2}\right]$

15 Evaluate (a) $\left(\dfrac{1}{3^2}\right)^{-1}$   (b) $81^{0.25}$   (c) $16^{-\frac{1}{4}}$   (d) $\left(\dfrac{4}{9}\right)^{\frac{1}{2}}$

$$\left[(a)\ 9 \ \ (b)\ \pm 3 \ \ (c)\ \tfrac{1}{2} \ \ (d)\ \pm\tfrac{2}{3}\right]$$

In *Problems 16 to 22*, evaluate the expressions given:

16 $\dfrac{9^2 \times 7^4}{3^4 \times 7^4 + 3^3 \times 7^2}$                                                $\left[\dfrac{147}{148}\right]$

17 $\dfrac{3^3 \times 5^2}{2^3 \times 3^2 - 8^2 \times 9}$                                     $\left[-1\dfrac{19}{56}\right]$

18 $\dfrac{3^3 \times 7^2 - 5^2 \times 7^3}{3^2 \times 5 \times 7^2}$                                  $\left[-3\dfrac{13}{45}\right]$

19 $\dfrac{(2^4)^2 - 3^{-2} \times 4^4}{2^3 \times 16^2}$                                     $\left[\dfrac{1}{9}\right]$

20 $\dfrac{\left(\dfrac{1}{2}\right)^3 - \left(\dfrac{2}{3}\right)^{-2}}{-\left(\dfrac{3}{5}\right)^2}$                                  $\left[-5\dfrac{65}{72}\right]$

21 $\left(\dfrac{4}{3}\right)^4 \Big/ \left(\dfrac{2}{9}\right)^2$                                       $[64]$

22 $\dfrac{(3^2)^{\frac{3}{2}} \times (8^{\frac{1}{3}})^2}{(3)^2 \times (4^3)^{\frac{1}{2}} \times (9)^{-\frac{1}{2}}}$                              $\left[4\dfrac{1}{2}\right]$

### Standard form

In *Problems 23 to 27*, express in standard form.

23 (a) 73.9;  (b) 28.4;  (c) 197.62.

$$[(a)\ 7.39 \times 10;\ (b)\ 2.84 \times 10;\ (c)\ 1.976\ 2 \times 10^2]$$

24 (a) 2 748;  (b) 33 170  (c) 274 218

$$[(a)\ 2.748 \times 10^3;\ (b)\ 3.317 \times 10^4;\ (c)\ 2.742\ 18 \times 10^5]$$

25 (a) 0.240 1;  (b) 0.017 4  (c) 0.009 23

$$[(a)\ 2.401 \times 10^{-1};\ (b)\ 1.74 \times 10^{-2};\ (c)\ 9.23 \times 10^{-3}]$$

26 (a) 1 702.3;  (b) 10.04  (c) 0.010 9

$$[(a)\ 1.702\ 3 \times 10^3;\ (b)\ 1.004 \times 10;\ (c)\ 1.09 \times 10^{-2}]$$

27 (a) $\dfrac{1}{2}$;  (b) $11\dfrac{7}{8}$;  (c) $130\dfrac{3}{5}$;  (d) $\dfrac{1}{32}$

$$[(a)\ 5 \times 10^{-1};\ (b)\ 1.187\ 5 \times 10;\ (c)\ 1.306 \times 10^2;\ (d)\ 3.125 \times 10^{-2}]$$

In *Problems 28 and 29*, express the numbers given as integers or decimal fractions.

28 (a) $1.01 \times 10^3$;  (b) $9.327 \times 10^2$;  (c) $5.41 \times 10^4$;  (d) $7 \times 10^0$

$$[(a)\ 1\ 010,\ (b)\ 932.7;\ (c)\ 54\ 100;\ (d)\ 7]$$

29 (a) $3.89 \times 10^{-2}$; (b) $6.741 \times 10^{-1}$; (c) $8 \times 10^{-3}$

[(a) 0.038 9; (b) 0.674 1; (c) 0.008]

In *Problems 30 to 33*, find the values of the expressions given, stating the answers in standard form.

30 (a) $3.7 \times 10^2 + 9.81 \times 10^2$; (b) $1.431 \times 10^{-1} + 7.3 \times 10^{-1}$;
(c) $2.68 \times 10^{-2} - 8.414 \times 10^{-2}$

[(a) $1.351 \times 10^3$; (b) $8.731 \times 10^{-1}$; (c) $-5.734 \times_{} 10^{-2}$]

31 (a) $4.831 \times 10^2 + 1.24 \times 10^3$; (b) $3.24 \times 10^{-3} - 1.11 \times 10^{-4}$;
(c) $1.81 \times 10^2 + 3.417 \times 10^2 - 5.972 \times 10^2$

[(a) 1.723 1 $\times 10^3$; (b) $3.129 \times 10^{-3}$; (c) $-7.45 \times 10$]

32 (a) $(4.5 \times 10^{-2})(3 \times 10^3)$; (b) $2 \times (5.5 \times 10^4)$

[(a) $1.35 \times 10^2$; (b) $1.1 \times 10^5$]

33 (a) $\dfrac{6 \times 10^{-3}}{3 \times 10^{-5}}$; (b) $\dfrac{(2.4 \times 10^3)(3 \times 10^{-2})}{(4.8 \times 10^4)}$

[(a) $2 \times 10^2$; (b) $1.5 \times 10^{-3}$]

34 Write the following statements in standard form.
(a) The density of aluminium is 2 710 kg m$^{-3}$
(b) Poisson's ratio for gold is 0.44
(c) The impedance of free space is 376.73 $\Omega$
(d) The electron rest energy is 0.511 Me V
(e) Proton charge–mass ratio is 95 789 700 C kg$^{-1}$
(f) The normal volume of a perfect gas is 0.022 41 m$^3$ mol$^{-1}$

[(a) $2.71 \times 10^3$ kg m$^{-3}$; (b) $4.4 \times 10^{-1}$; (c) $3.767\ 3 \times 10^2$ $\Omega$;
(d) $5.11 \times 10^{-1}$ Me V; (e) $9.578\ 97 \times 10^7$ C kg$^{-1}$;
(f) $2.241 \times 10^{-2}$ m$^3$ mol$^{-1}$]

*Binary form*

In *Problems 35 and 36*, convert the denary numbers given to binary numbers.

35 (a) 5; (b) 15; (c) 17; (d) 29.

[(a) $101_2$; (b) $1111_2$; (c) $10001_2$; (d) $11101_2$]

36 (a) 31; (b) 42; (c) 57; (d) 63.

[(a) $11111_2$; (b) $101\ 010_2$; (c) $111\ 001_2$; (d) $111\ 111_2$]

In *Problems 37 and 38*, convert the binary numbers given to denary numbers.

37 (a) 110; (b) 1011; (c) 1110; (d) 1001

[(a) $6_{10}$; (b) $11_{10}$; (c) $14_{10}$; (d) $9_{10}$]

38 (a) 10101; (b) 11001; (c) 101 101; (d) 110 011

[(a) $21_{10}$; (b) $25_{10}$; (c) $45_{10}$; (d) $51_{10}$]

In *Problems 39 to 41*, add the binary numbers given, giving the answers in both binary and denary forms.

39 (a) 101+100; (b) 111+1011; (c) 1010+1101

[(a) $1001_2$, $9_{10}$; (b) $10010_2$; $18_{10}$; (c) $10111_2$, $23_{10}$]

40 (a) 1110+1011; (b) 1010+1011; (c) 1111+1111

[(a) $11001_2$; $25_{10}$; (b) $10101_2$; $21_{10}$; (c) $11110_2$; $30_{10}$]

41 (a) $10111 + 10010$; (b) $11010 + 10110$; (c) $1011 + 101\ 110$

[(a) $101\ 001_2$; $41_{10}$; (b) $110\ 000_2$; $48_{10}$; (c) $111\ 001_2$; $57_{10}$]

In *Problems 42 to 44*, add the binary numbers given. Add the first two numbers and then add the third number to the result of the first addition. Check the answers by converting each binary number to denary form.

42  $1000 + 10001 + 100\ 010$                    [$111\ 011_2$, $8 + 17 + 34 = 59_{10}$]

43  $1011 + 10010 + 10111$                    [$110\ 100$, $11 + 18 + 23 = 52_{10}$]

44  $100\ 101 + 10001 + 1001$                    [$111\ 111_2$, $37 + 17 + 9 = 63_{10}$]

# 3 Calculations

## A. MAIN POINTS INVOLVED WITH ARITHMETIC CALCULATIONS

1. (i) In all problems in which the measurement of distance, time, mass or other quantities occurs, an exact answer cannot be given; only an answer which is correct to a stated degree of accuracy can be given. To take account of this an **error due to measurement** is said to exist.
   (ii) To take account of measurement errors it is usual to limit answers so that the result given is **not more than one significant figure greater than the least accurate number given in the data.**
   (iii) **Rounding-off errors** can exist with decimal fractions. For example, to state that $\pi = 3.142$ is not strictly correct, but '$\pi = 3.142$ correct to 4 significant figures' is a true statement. (Actually, $\pi = 3.14159265 \ldots\ldots$ .)
   (iv) It is possible, through an incorrect procedure, to obtain the wrong answer to a calculation. This type of error is known as a **blunder**.
   (v) An **order of magnitude error** is said to exist if incorrect positioning of the decimal point occurs after a calculation has been completed.
   (vi) Blunders and order of magnitude errors can be reduced by determining **approximate values of calculations**. Answers which do not seem feasible must be checked and the calculation must be repeated as necessary. (see *Problems 1 and 2*)

2. Aids to arithmetic calculations include (i) 4 figure tables of squares, square roots and reciprocals, (ii) 4 figure tables of common logarithms and antilogarithms, (iii) slide rules, and (iv) calculators.

3. With **4-figure tables** the results stated are correct to 3 significant figures, the fourth significant figure being correct to $\pm 1$.
   (i) **Square tables** give the squares of numbers from 1.0 to 9.999.
      *For example*
      $4.2^2 = 17.64$
      $4.26^2 = 18.15$
      $4.265^2 = 18.15 + 4$ (from the mean difference column corresponding to 5)
      i.e. $4.265^2 = 18.19$

      Similarly
      $1.287^2 = 1.655$ and $8.973^2 = 80.51$
      For the square of a number outside of the range 1.0 to 9.999, the number should initially be expressed in standard form. Thus

      $31.73^2 = (3.173 \times 10)^2 = 3.173^2 \times 10^2 = 10.07 \times 10^2 = 1\,007$ and
      $0.06815^2 = (6.815 \times 10^{-2})^2 = 6.815^2 \times 10^{-4} = 46.45 \times 10^{-4}$
      $= 0.004\,645$

      (See *Problem 3*)

(ii) **Square root tables** give the square roots of numbers from 1.0 to 99.99.

*For example*

$\sqrt{3.4}$ $= \pm 1.844$ (note that a square root always has two answers)

$\sqrt{3.47} = \pm 1.863$

$\sqrt{3.476} = \pm 1.863 + 2$ (from the mean difference column corresponding to 6)

i.e. $\sqrt{3.476} = \pm 1.865$

Similarly

$\sqrt{8.738} = \pm 2.956$ and $\sqrt{65.93} = \pm 8.120$

For the square root of a number outside of the range 1.0 and 99.99, the number should be expressed as a number between 1.0 and 99.99 multiplied by 10 raised to an even power. Thus

$$\sqrt{437} = (437)^{1/2} = (4.37 \times 10^2)^{1/2} = 4.37^{1/2} \times (10^2)^{1/2} = \sqrt{4.37} \times 10$$
$$= \pm 2.090 \times 10 = \pm 20.90$$
$$\text{and } \sqrt{0.007\ 631} = (76.31 \times 10^{-4})^{1/2} = (76.31)^{1/2} \times (10^{-4})^{1/2} = \sqrt{76.31} \times 10^{-2}$$
$$= \pm 8.735 \times 10^{-2} = \pm 0.087\ 35$$

(See *Problem 4*)

(iii) **Reciprocal tables** give reciprocals of numbers from 1.0 to 9.999.

*For example*

$\dfrac{1}{4.8} = 0.2083$

$\dfrac{1}{4.87} = 0.2053$

$\dfrac{1}{4.879} = 0.2053 - 4$ (from the mean difference column corresponding to 9)

i.e. $\dfrac{1}{4.879} = 0.2049$

Note, for reciprocals, the numbers in the difference columns are **subtracted**. not added.

Similarly $\dfrac{1}{2.364} = 0.4230$ and $\dfrac{1}{9.718} = 0.1029$

For the reciprocal of a number outside of the range 1.0 to 9.999, the number should initially be expressed in standard form. Thus

$$\frac{1}{32.68} = \frac{1}{3.268 \times 10} = \frac{1}{3.268} \times 10^{-1} = 0.3059 \times 10^{-1} = 0.03059$$

and

$$\frac{1}{0.00529} = \frac{1}{5.29 \times 10^{-3}} = \frac{1}{5.29} \times 10^{+3} = 0.1890 \times 10^3 = 189.0$$

(See *Problem 5*)

4 **Common logarithms** may be used when multiplying or dividing numbers or for raising numbers to powers.

(i) **Definition of a logarithm** If $y = a^x$ then $x = \log_a y$.

'The common logarithm of a number is the power to which 10 has to be raised to be equal to the number.' Thus, since $100 = 10^2$, then 2 is the logarithm of 100 to a base 10 and is written as $2 = \log_{10} 100$, or more usually, $2 = \lg 100$ (where lg means $\log_{10}$).

## (ii) Tables of logarithms

A logarithm is made up of two parts: (a) an integer called the **characteristic**, and (b) a decimal part called the **mantissa** (which is always positive). To find the logarithm of any positive number, initially change the number to standard form. The power to which the 10 is raised gives the characteristic and the mantissa is read from logarithm tables in a similar manner to that used for squares and square roots.

*For example*

$\lg 3.564$ $= \lg (3.564 \times 10^0) = 0.5519$
$\lg 356.4$ $= \lg (3.564 \times 10^2) = 2.5519$
$\lg 0.003564$ $= \lg (3.564 \times 10^{-3}) = -3 + 0.5519$, which is written as $\bar{3}.5519$.

Similarly $\lg 8672 = 3.9381$, $\lg 0.127 = \bar{1}.1038$ and $\log 1 = \lg 10^0 = 0$.

## (iii) Tables of antilogarithms

The reverse process to finding the logarithm of a number is to find the antilogarithm of a number which can be determined by using tables. Only the mantissa is used with antilogarithm tables; the characteristic merely fixes the position of the decimal point. For example, to find the antilogarithm of 3.9381, .9381 in antilogarithmic tables corresponds to 8.672 (note that antilogarithm tables always give a four-figure number which is then changed into a number between 1.0 and 9.999). The characteristic 3 indicates that 8.672 is multiplied by $10^3$.

Hence, the antilogarithm of 3.9381 is 8672.

Similarly, if $\lg N = 1.7214$, $N = 5.265 \times 10^1 = 52.65$,
and if $\lg N = \bar{2}.3689$, $N = 2.338 \times 10^{-2} = 0.02338$.

## (iv) Rules of logarithms

(a) To multiply two numbers, $\lg (A \times B) = \lg A + \lg B$.

(b) To divide two numbers, $\lg \left( \dfrac{A}{B} \right) = \lg A - \lg B$.

(c) To raise a number to a power, $\lg A^n = n \times \lg A$.

(See *Problems 6 to 18*)

5  A **slide rule** provides a method of performing calculations quickly, correct to about 3 significant figure accuracy. However, their use, together with the use of tables, has declined with the advent of **electronic calculators**. The scientific notation type calculator can perform all the functions provided by tables and slide rules more quickly and with greater accuracy (8 digit displays are available on most calculators). However answers should be limited to a reasonable number of significant figures (see para 1(ii)).

There are many varying makes of calculators and slide rules and each usually have detailed instructions for their use supplied at purchase. 4-figure tables should not, however, be completely abandoned since they remain the cheapest of all aids to calculations and are not dependent upon a source of electrical power. (See *Problem 19*)

## B. WORKED PROBLEMS INVOLVING ARITHMETIC CALCULATIONS

**Problem 1** The area $A$ of a triangle is given by $A = \frac{1}{2}bh$. The base $b$ when measured is found to be 3.26 cm, and the perpendicular height $h$ is 7.5 cm. Determine the area of the triangle.

Area of triangle = $\frac{1}{2}bh = \frac{1}{2} \times 3.26 \times 7.5 = 12.225$ cm$^2$ (by long multiplication or by calculator).

The approximate value is $\frac{1}{2} \times 3 \times 8 = 12$ cm$^2$, so there are no obvious blunder or of magnitude errors. However it is not usual in a measurements type problem to state the answer to an accuracy greater than 1 significant figure more than the least accurate number in the data; this is 7.5 cm, so the result should not have more than 3 significant figures.

**Thus area of triangle = 12.2 cm$^2$**

**Problem 2** State which type of error has been made in the following statements:
(a) $72 \times 31.429 = 2\,262.9$, (b) $16 \times 0.08 \times 7 = 89.6$,
(c) $11.714 \times 0.0088 = 0.3247$ correct to 4 decimal places,

(d) $\dfrac{29.74 \times 0.0512}{11.89} = 0.12$, correct to 2 significant figures.

(a) $72 \times 31.429 = 2\,262.888$ (by long multiplication or calculator), hence a **rounding-off error** has occurred. The answer should have stated:
$72 \times 31.429 = 2\,262.9$ correct to 5 significant figures.

(b) $16 \times 0.08 \times 7 = \overset{4}{\cancel{16}} \times \dfrac{8}{\underset{25}{\cancel{100}}} \times 7 = \dfrac{32 \times 7}{25} = \dfrac{224}{25} = 8\dfrac{24}{25} = 8.96$

Hence an **order of magnitude error** has occurred.

(c) $11.714 \times 0.0088$ is approximately equal to $12 \times 9 \times 10^{-3}$, i.e. about $108 \times 10^{-3}$ or 0.108. Thus a **blunder** has been made.

(d) $\dfrac{29.74 \times 0.0512}{11.89} \simeq \dfrac{30 \times 5 \times 10^{-2}}{12} \simeq \dfrac{150}{12 \times 10^2} \simeq \dfrac{15}{120} \simeq \dfrac{1}{8}$ or 0.125,

hence no order of magnitude error has occurred. However,

$\dfrac{29.74 \times 0.0512}{11.89} = 0.128$ correct to 3 significant figures, which equals 0.13 correct to 2 significant figures. Hence a **rounding-off error** has occurred.

**Problem 3** Use 4-figure tables to find the values of (a) $6.847^2$; (b) $246.1^2$; (c) $0.00451^2$

(a) $6.8^2$ = 46.24
$6.84^2$ = 46.79
$6.847^2$ = 46.79+10 (from the mean difference column corresponding to 7)
i.e. $6.847^2$ = **46.89**
(b) $246.1^2 = (2.461 \times 10^2)^2 = 2.461^2 \times 10^4 = 6.057 \times 10^4$ = **60570**
(c) $0.00451^2 = (4.51 \times 10^{-3})^2 = 4.51^2 \times 10^{-6} = 20.34 \times 10^{-6}$ = **0.00002034**

---

*Problem 4* Use 4-figure tables to find the values of (a) $\sqrt{2.683}$; (b) $\sqrt{78.65}$; (c) $\sqrt{56230}$; (d) $\sqrt{0.001796}$

---

(a) $\sqrt{2.6}$ = ±1.612
$\sqrt{2.68}$ = ±1.637
$\sqrt{2.683}$ = ±1.637+1 (from the mean difference column corresponding to 3)
i.e. $\sqrt{2.683}$ = **±1.638**
(b) $\sqrt{78.65}$ = **±8.869**
(c) $\sqrt{56230} = \sqrt{(5.623 \times 10^4)} = \sqrt{5.623} \times \sqrt{10^4} = \sqrt{5.623} \times 10^2$
$= \pm 2.372 \times 10^2$ = **±237.2**
(d) $\sqrt{0.001796} = \sqrt{(17.96 \times 10^{-4})} = \sqrt{17.96} \times \sqrt{10^{-4}} = \sqrt{17.96} \times 10^{-2}$
$= \pm 4.238 \times 10^{-2}$ = **±0.04238**

---

*Problem 5* Use 4-figure tables to find the values of (a) $\dfrac{1}{3.467}$; (b) $\dfrac{1}{672.3}$; (c) $\dfrac{1}{0.007815}$; (d) $\dfrac{1}{1.138}$

---

(a) $\dfrac{1}{3.4}$ = 0.2941

$\dfrac{1}{3.46}$ = 0.2890

$\dfrac{1}{3.467}$ = 0.2890−6 (from the mean difference column corresponding to 7)

i.e. $\dfrac{1}{3.467}$ = **0.2884** (Note that for reciprocals, the numbers in the difference columns are subtracted.)

(b) $\dfrac{1}{672.3} = \dfrac{1}{6.723 \times 10^2} = \dfrac{1}{6.723} \times 10^{-2} = 0.1487 \times 10^{-2}$ = **0.001487**

(c) $\dfrac{1}{0.007815} = \dfrac{1}{7.815 \times 10^{-3}} = \dfrac{1}{7.815} \times 10^3 = 0.1279 \times 10^3$ = **127.9**

(d) No mean differences are given in the region of 1.138, thus the number must be rounded off to 1.14 and

$\dfrac{1}{1.14}$ = **0.8772**

(a) lg 5.279 = lg (5.279 × $10^0$)
The characteristic (i.e. integer) is given by the power of 10, in this case, 0.
The mantissa (i.e. decimal part) is read from logarithm tables.
Hence lg (5.279 × $10^0$) = **0.7225**
(b) lg 378.2 = lg (3.782 × $10^2$) = **2.5777**
(c) lg 0.004 831 = lg (4.831 × $10^{-3}$) = −3+0.6840 = $\overline{3}$**.6840**

(a) Only the decimal part is used when reading antilogarithm tables. 0.2783 from antilogarithm tables corresponds to 1.898 (i.e. a number between 1.0 and 9.999). The characteristic 0 indicates that 1.898 is multiplied by $10^0$, (i.e. 1). Hence the antilogarithm of 0.2783 is **1.898**
(b) From antilogarithm tables, 0.8329 corresponds to 6.806. The characteristic 3 indicates that 6.806 is multiplied by $10^3$. Hence the antilogarithm of 3.8329 is 6.806 × $10^3$, i.e. **6806**
(c) The antilogarithm of 1.6377 is 4.342 × $10^{-1}$, i.e. **0.4342**

For multiplication, the logarithms are added, i.e. lg (A × B) = lg A+lg B.

| Number | Logarithm |
|---|---|
| 26.47 = 2.647 × $10^1$ | 1.4227 |
| 0.02981 = 2.981 × $10^{-2}$ | $\overline{2}$.4743 |
| 7.889 × $10^{-1}$ | $\overline{1}$.8970  Adding |

Hence 26.47 × 0.02981 = 7.889 × $10^{-1}$ = **0.7889**

For multiplication, the logarithms are added, no matter how many numbers involved.

| Number | Logarithm |
|---|---|
| $46.32 = 4.632 \times 10^1$ | 1.6658 |
| $97.17 = 9.717 \times 10^1$ | 1.9875 |
| $0.1258 = 1.258 \times 10^{-1}$ | $\overline{1}.0996$ |
| $5.661 \times 10^2$ | 2.7529 Adding |

Hence $46.32 \times 97.17 \times 0.1258$
$= 5.661 \times 10^2 = \mathbf{566.1}$

*Problem 10* Use logarithms to evaluate $\dfrac{4.621}{23.76}$

When numbers are divided, the logarithms are subtracted, i.e. $\lg (A/B) = \lg A - \lg B$

| Number | Logarithm |
|---|---|
| $4.621 = 4.621 \times 10^0$ | 0.664 7 |
| $23.76 = 2.376 \times 10^1$ | 1.375 8 |
| $1.945 \times 10^{-1}$ | $\overline{1}.288\ 9$ Subtracting |

Hence $\dfrac{4.621}{23.76}$
$= 1.945 \times 10^{-1} = \mathbf{0.1945}$

*Problem 11* Use logarithms to evaluate 1/52.73

| Number | Logarithm |
|---|---|
| $1 = 1 \times 10^0$ | $\overline{1}.\cancel{0}^9\cancel{0}^9\cancel{0}^{10}0$ |
| $52.73 = 5.273 \times 10^1$ | 1.7 2 2 0 |
| $1.897 \times 10^{-2}$ | $\overline{2}.2\ 7\ 8\ 0$ Subtracting |

(Note, if 1 is 'borrowed' from 0 then $\overline{1}$ is left. Alternatively, add 1 to top and bottom)

Hence $\dfrac{1}{52.73} = 1.897 \times 10^{-2} = \mathbf{0.01897}$ (which may be checked using reciprocal tables)

*Problem 12* Use logarithms to evaluate $(31.28)^3$

To find the *n*'th power of a number, the logarithm of the number is multiplied by n, i.e. $\lg A^n = n \lg A$.

| Number | Logarithm |
|---|---|
| $31.28 = 3.128 \times 10^1$ | 1.495 3 |
| | 3 |
| $3.061 \times 10^4$ | 4.485 9 Multiplying |

Hence $(31.28)^3 = 3.061 \times 10^4$
$= \mathbf{30610}$

**Problem 13** Evaluate, using logarithms $(0.8316)^4$

| Number | Logarithm |
|--------|-----------|
| $0.8316 = 8.316 \times 10^{-1}$ | $\overline{1}.919\ 9$ |
| | 4 |
| $4.782 \times 10^{-1}$ | $\overline{1}.679\ 6$ Multiplying |

Hence $(0.8316)^4$
$= 4.782 \times 10^{-1}$
$= \mathbf{0.4782}$

---

**Problem 14** Determine $\sqrt{546.2}$ using logarithms

| Number | Logarithm |
|--------|-----------|
| $546.2 = 5.462 \times 10^2$ | 2 ) 2.7374 |
| $2.337 \times 10^1$ | 1.3687 |

Since $\sqrt{546.2} = (546.2)^{1/2}$ then the logarithm of 546.2 is multiplied by $\frac{1}{2}$, which is the same as dividing by 2.
Hence $\sqrt{546.2}$
$= \pm 2.337 \times 10^1 = \mathbf{\pm 23.37}$

---

**Problem 15** Find the value of $\sqrt{0.007\ 328}$ using logarithms

| Number | Logarithm |
|--------|-----------|
| $0.007\ 328 = 7.328 \times 10^{-3}$ | $\overline{3}.8650$ |
| | $2)\overline{4}+1.8650$ |
| $8.561 \times 10^{-2}$ | $\overline{2}+0.9325$ |

If the characteristic $\overline{3}$ is divided by 2 it does not give an exact integer answer. Hence $-1$ is added to $\overline{3}$ to make $\overline{4}$ and 1 is added to the mantissa to make the overall value of the logarithm the same, i.e. $\overline{3}.865\ 0$ is the same as $\overline{4}+1.865\ 0$

Hence $\sqrt{0.007\ 328} = 8.561 \times 10^{-2} = \mathbf{0.085\ 61}$

---

**Problem 16** Evaluate $\left(\dfrac{19.74 \times 0.086\ 1}{3.462}\right)^4$ using logarithms

| Number | Logarithm |
|--------|-----------|
| $19.74 = 1.974 \times 10^1$ | $1.2\ 9\ 5\ \ 3$ |
| $0.0861 = 8.61 \times 10^{-2}$ | $\overline{2}.9\ 3\ 5\ \ 0$ |
| | $\overline{1}.2\ 3\ 0\ \ 3$ 13 Adding |
| $3.462 = 3.462 \times 10^0$ | $0.5\ 3\ 9\ \ 4$ |
| | $\overline{1}.6\ 9\ 0\ \ 9$ Subtracting |
| | 4 |
| $5.802 \times 10^{-2}$ | $\overline{2}.7\ 6\ 3\ \ 6$ Multiplying |

Hence $\left(\dfrac{19.74 \times 0.0861}{3.462}\right)^4$
$= 5.802 \times 10^{-2}$
$= \mathbf{0.058\ 02}$

*Problem 17* Use logarithms to find the energy $E$ where $E = \frac{1}{2}LI^2$ joules given that current $I = 4.78$ A and inductance $L = 0.342$ H

$E = \frac{1}{2}LI^2 = \frac{1}{2} \times 0.342 \times 4.78^2$

| Number | Logarithm |
|--------|-----------|
| $I = 4.78 = 4.78 \times 10^0$ | 0.679 4 |
| | 2 |
| $I^2$ | 1.358 8 Multiplying |
| $L = 0.342 = 3.42 \times 10^{-1}$ | 1.534 0 |
| $LI^2 = 7.812 \times 10^0$ | 0.892 8 Adding |

Hence $E = \frac{1}{2}LI^2 = \frac{1}{2} \times 7.812 \times 10^0 = \dfrac{7.812}{2} = \mathbf{3.906 \ J}$

*Problem 18* Kinetic energy $E$ is given by $E = \frac{1}{2}m(v^2 - u^2)$. Use logarithms to evaluate $E$ when mass $m = 220$ kg, final velocity $v = 72.3$ m/s and the initial velocity $u = 33.7$ m/s.

Addition and subtraction **cannot** be achieved using logarithms. $v^2$ and $u^2$ may be determined using logarithms, but the value of $(v^2 - u^2)$ must be determined by subtraction.

| Number | Logarithm |
|--------|-----------|
| $v = 72.3 = 7.23 \times 10^1$ | 1.859 1 |
| | 2 |
| $v^2 = 522 6$ | 3.718 2 |

| Number | Logarithm |
|--------|-----------|
| $u = 33.7 = 3.37 \times 10^1$ | 1.527 6 |
| | 2 |
| $u^2 = 1136$ | 3.055 2 |

Thus $v^2 - u^2 = 522\ 6 - 113\ 6 = \mathbf{4\ 090}$

| Number | Logarithm |
|---|---|
| $v^2 - u^2 = 4090 = 4.090 \times 10^3$ | 3.611 7 |
| $m = 220 = 2.20 \times 10^2$ | 2.342 4 |
| $m(v^2 - u^2) = 8.997 \times 10^5$ | 5.954 1 |

Thus $\frac{1}{2}m(v^2 - u^2) = \frac{8.997 \times 10^5}{2} = \frac{899\ 700}{2} = $ **449 850 J**

From the data given the answer should be expressed to no more than 4 significant figures. **Thus kinetic energy $E = $ 449 900 J**

*Problem 19* Using a calculator and/or slide rule check the answers to *Problems 1 to 18* and compare the accuracy of each method.

## C. FURTHER PROBLEMS INVOLVING ARITHMETIC CALCULATIONS

1   The area $A$ of a rectangle is given by $A = lb$. The length $l$ when measured is found to be 23.1 mm and the breadth $b$ is 7.8 mm. Determine the area of the rectangle.

$[180\ \text{mm}^2]$

2   The velocity of a body is given by $v = u + at$. The initial velocity $u$ is measured when time $t$ is 15 seconds and found to be 12 m/s. If the acceleration $a$ is 9.81 m/s$^2$ calculate the final velocity $v$.

$[159\ \text{m/s}]$

3   Calculate the current $I$ in an electrical circuit, where $I = V/R$ amperes when the voltage $V$ is measured and found to be 7.2 V and the resistance $R$ is 17.7 $\Omega$.

$[0.407\ \text{A}]$

In problems 4 to 8 state which type of error, or errors, have been made.

4   $25 \times 0.06 \times 1.4 = 0.21$     [order of magnitude error]
5   $137 \times 6.842 = 937.4$     $\begin{bmatrix} \text{rounding-off error—should add} \\ \text{'correct to 4 significant figures'} \end{bmatrix}$

6   $\dfrac{204 \times 0.008}{12.6} = 10.42$     [Blunder]

7   For a gas $pV = c$. When pressure $p = 103\ 400$ Pa and $V = 0.54$ m$^3$ then $c = 55\ 836$ Pa m$^3$

[Measured values, hence $c = 55\ 800$ Pa m$^3$]

8   $\dfrac{4.6 \times 0.07}{52.3 \times 0.274} = 0.225$     $\begin{bmatrix} \text{order of magnitude error and rounding-off} \\ \text{error—should be 0.0225 correct to 3} \\ \text{significant figures} \end{bmatrix}$

In *Problems 9 to 13*, use 4-figure tables to evaluate the quantities shown.

9   (a) $3.249^2$; (b) $73.78^2$; (c) $311.4^2$; (d) $0.0639^2$

[(a) 10.56; (b) 5444; (c) 96 970; (d) 0.004 083]

10 (a) $\sqrt{4.735}$; (b) $\sqrt{35.46}$; (c) $\sqrt{73\ 280}$; (d) $\sqrt{0.025\ 6}$

$\qquad$ [(a) 2.176; (b) 5.955; (c) 270.7; (d) 0.160]

11 (a) $\dfrac{1}{7.768}$ ; (b) $\dfrac{1}{48.46}$ ; (c) $\dfrac{1}{0.0816}$ ; (d) $\dfrac{1}{1.118}$

$\qquad$ [(a) 0.128 8; (b) 0.020 63; (c) 12.25; (d) 0.892 9]

12 (a) lg 3.764; (b) lg 241.8; (c) lg 1.0; (d) lg 0.076 32

$\qquad$ [(a) 0.575 7; (b) 2.383 4; (c) 0; (d) $\overline{2}$.882 6]

13 (a) antilogarithm of 0.586 2; (b) antilogarithm of 4.731,
  (c) antilogarithm of $\overline{2}$.319 7 $\qquad$ [(a) 3.857; (b) 53 830; (c) 0.020 87]

In *Problems 14 to 30*, use 4-figure logarithms to evaluate.

14 (a) 43.27 × 12.91 $\qquad$ (b) 54.31 × 0.572 4

$\qquad$ [(a) 558.6; (b) 31.09]

15 (a) 127.8 × 0.043 1 × 19.8 $\quad$ (b) 15.76 ÷ 4.329

$\qquad$ [(a) 109.1; (b) 3.641]

16 (a) $\dfrac{137.6}{552.9}$ $\qquad$ (b) $\dfrac{11.82 \times 1.736}{0.041}$

$\qquad$ [(a) 0.248 9; (b) 500.3]

17 (a) $\dfrac{1}{17.31}$ $\qquad$ (b) $\dfrac{1}{0.0346}$ $\qquad$ (c) $\dfrac{1}{147.9}$

$\qquad$ [(a) 0.057 77; (b) 28.90; (c) 0.006 763]

18 (a) $13.6^3$ $\qquad$ (b) $3.476^4$ $\qquad$ (c) $0.124^5$

$\qquad$ [(a) 2 515; (b) 146.0; (c) 0.000 029 31]

19 (a) $\sqrt{347.1}$ (b) $\sqrt{7\ 632}$ (c) $\sqrt{0.027}$ (d) $\sqrt{0.004\ 168}$

$\qquad$ [(a) 18.63; (b) 87.36; (c) 0.164 4; (d) 0.064 55]

20 (a) $\left(\dfrac{24.68 \times 0.053\ 2}{7.412}\right)^3$ $\qquad$ (b) $\left(\dfrac{0.2681 \times 41.2^2}{32.6 \times 11.89}\right)^4$

$\qquad$ [(a) 0.005 558; (b) 1.900]

21 (a) $\dfrac{14.32^3}{21.68^2}$ $\qquad$ (b) $\dfrac{4.821^3}{17.33^2 - 15.86 \times 11.6}$

$\qquad$ [(a) 6.244; (b) 0.966 8]

22 Find the distance $s$, given that $s = \frac{1}{2}gt^2$. Time $t = 0.032$ seconds and acceleration due to gravity $g = 9.81$ m/s$^2$.

$\qquad$ [0.005 02 m or 5.02 mm]

23 The energy stored in a capacitor is given by $E = \frac{1}{2}CV^2$ joules. Determine the energy when capacitance $C = 5 \times 10^{-6}$ farads and voltage $V = 240$ V.

$\qquad$ [0.144 0 J]

24 Find the area $A$ of a triangle, given $A = \frac{1}{2}bh$, when the base length $l$ is 23.42 m and the height $h$ is 53.7 m. $\qquad$ [629.0 m$^2$]

25 Resistance $R_2$ is given by $R_2 = R_1(1+\alpha t)$. Find $R_2$, correct to 4 significant figures when $R_1 = 220$, $\alpha = 0.000\ 27$ and $t = 75.6$.

$\qquad$ [224.5]

26 Density $= \dfrac{\text{mass}}{\text{volume}}$ . Find the density when the mass is 2.462 kg and the volume is 173 cm$^3$. Give the answer in units of kg/m$^3$.

$\qquad$ [14 230 kg/m$^3$]

27  Velocity = frequency × wavelength. Find the velocity when the frequency is 1825 Hz and the wavelength is 0.154 m.

$$[281.0 \text{ m/s}]$$

28  Evaluate resistance $R_T$, given $\dfrac{1}{R_T} = \dfrac{1}{R_1} + \dfrac{1}{R_2} + \dfrac{1}{R_3}$ , when
$R_1 = 5.5 \ \Omega$, $R_2 = 7.42 \ \Omega$ and $R_3 = 12.6 \ \Omega$.

$$[2.526 \ \Omega]$$

29  Find the total cost of 37 calculators costing £12.65 each and 19 slide rules costing £6.38 each.

$$[£589.30]$$

30  Power = $\dfrac{\text{force} \times \text{distance}}{\text{time}}$. Find the power when a force of 3760 N raises an object a distance of 4.73 m in 35 s.

$$[508.2 \text{ W}]$$

31  In problems 1 to 30, evaluate using a calculator and/or slide rule.

32  Currency exchange rates quoted from a newspaper on 8 July 1980 included the following:

| France | £1 = 9.41 francs |
|---|---|
| Japan | £1 = 515.0 yen |
| West Germany | £1 = 4.06 Deutschmarks |
| U.S.A. | £1 = $2.3475 |
| Spain | £1 = 160.0 pesetas |

Calculate (a) how many French francs £32.50 will buy, (b) the number of American dollars that can be purchased for £74.80, (c) the pounds sterling which can be exchanged for 14 000 yen, (d) the pounds sterling which can be exchanged for 1750 pesetas, and (e) the West German Deutschmarks which can be bought for £55.

$$\left[ \begin{array}{l} \text{(a) 305.83 f; (b) \$175.59; (c) £27.18;} \\ \text{(d) £10.94; (e) 223.3 Dm} \end{array} \right]$$

33  Below is a table of some metric to imperial conversions.

| Length | 2.54 cm = 1 inch |
|---|---|
| | 1.61 km = 1 mile |
| Weight | 1 kg = 2.2 lb (1 lb = 16 ounces) |
| Capacity | 1 litre = 1.76 pints (8 pints = 1 gallon) |

Use the table to determine (a) the number of millimetres in 15 inches, (b) a speed of 35 mph in km/h, (c) the number of kilometres in 235 miles, (d) the number of pounds and ounces in 24 kg (correct to the nearest ounce), (e) the number of kilograms in 15 lb, (f) the number of litres in 12 gallons, and (g) the number of gallons in 25 litres.

$$\left[ \begin{array}{l} \text{(a) 381 mm; (b) 56.35 km/h; (c) 376 km; (d) 52 lbs 13 ozs;} \\ \text{(e) 6.82 kg; (f) 54.55 l; (g) 5.5 gallons.} \end{array} \right]$$

# 4 Introduction to algebra

## A. MAIN POINTS CONCERNED WITH ALGEBRA

1 **Algebra** is that part of mathematics in which the relations and properties of numbers are investigated by means of general symbols. For example, the area of a rectangle is found by multiplying the length by the breadth; this is expressed algebraically as $A = l \times b$, where $A$ represents the area, $l$ the length and $b$ the breadth.

2 The basic laws introduced in arithmetic are generalised in algebra. Let $a, b, c$ and $d$ represent any four numbers. Then:

(i) $a + (b+c) = (a+b) + c$

(ii) $a(bc) = (ab)c$

(iii) $a+b = b+a$

(iv) $ab = ba$

(v) $a(b+c) = ab+ac$

(vi) $\dfrac{a+b}{c} = \dfrac{a}{c} + \dfrac{b}{c}$

(vii) $(a+b)(c+d) = ac+ad+bc+bd$

3 **Laws of indices:**

(i) $a^m \times a^n = a^{m+n}$

(ii) $\dfrac{a^m}{a^n} = a^{m-n}$

(iii) $(a^m)^n = a^{mn}$

(iv) $a^{\frac{m}{n}} = \sqrt[n]{a^m}$

(v) $a^{-n} = \dfrac{1}{a^n}$

(vi) $a^0 = 1$

4 When two or more terms in an algebraic expression contain a common factor, then this factor can be shown outside of a bracket. For example

$ab+ac = a(b+c)$, which is simply the reverse of (v) in para 2,
and
$6px+2py-4pz = 2p(3x+y-2z)$
This process is called **factorisation**.

5 The **laws of precedence** which apply to arithmetic also apply to algebraic expressions. The order is *B*rackets, *O*f, *D*ivision, *M*ultiplication, *A*ddition and *S*ubtraction (i.e. BODMAS).

46

**Direct and inverse proportionality**

6　An expression such as $y = 3x$ contains two variables. For every value of $x$ there is a corresponding value of $y$. The variable $x$ is called the **independent variable** and $y$ is called the **dependent variable**.

7　When an increase or decrease in an independent variable leads to an increase or decrease of the same proportion in the dependent variable this is termed **direct proportion**. If $y = 3x$ then $y$ is directly proportional to $x$, which may be written as $y \propto x$ or $y = kx$, where $k$ is called the **coefficient of proportionality** (in this case, $k$ being equal to 3).

8　When an increase in an independent variable leads to a decrease of the same proportion in the dependent variable (or vice versa) this is termed **inverse proportion**. If $y$ is inversely proportional to $x$ then $y \propto (1/x)$ or $y = k/x$. Alternatively, $k = xy$, that is, for inverse proportionality the product of the variables is constant.

9　Examples of laws involving direct and inverse proportion in science include:

　(i) **Hookes law**, which states that within the elastic limit of a material, the strain $\epsilon$ produced is directly proportional to the stress, $\sigma$, producing it, i.e. $\epsilon \propto \sigma$ or $\epsilon = k\sigma$.

　(ii) **Charle's law**, which states that for a given mass of gas at constant pressure the volume $V$ is directly proportional to its thermodynamic temperature $T$, i.e. $V \propto T$ or $V = kT$.

　(iii)**Ohm's law**, which states that the current $I$ flowing through a fixed resistor is directly proportional to the applied voltage $V$, i.e. $I \propto V$ or $I = kV$.

　(iv) **Boyle's law**, which states that for a gas at constant temperature, the volume $V$ of a fixed mass of gas is inversely proportional to its absolute pressure $p$, i.e. $p \propto (1/V)$ or $p = k/V$, i.e. $pV = k$. (See *Problems 50 to 53*.)

## B. WORKED PROBLEMS ON ALGEBRA

(a) BASIC OPERATIONS

*Problem 1*　Evaluate $3ab - 2bc + abc$ when $a = 1$, $b = 3$ and $c = 5$

Replacing $a$, $b$, and $c$ with their numerical values gives:

$$3ab - 2bc + abc = 3 \times 1 \times 3 - 2 \times 3 \times 5 + 1 \times 3 \times 5 = 9 - 30 + 15$$
$$= -6$$

*Problem 2*　Find the value of $4p^2qr^3$, given that $p = 2$, $q = \frac{1}{2}$ and $r = 1\frac{1}{2}$

Replacing $p$, $q$ and $r$ with their numerical values gives:

$$4p^2qr^3 = 4(2)^2 \left(\frac{1}{2}\right) \left(\frac{3}{2}\right)^3 = 4 \times 2 \times 2 \times \frac{1}{2} \times \frac{3}{2} \times \frac{3}{2} \times \frac{3}{2} = 27$$

The sum of the positive terms is $3x+2x = 5x$
The sum of the negative terms is $x+7x = 8x$
Taking the sum of the negative terms from the sum of the positive terms gives:

$5x-8x = -3x$

Alternatively, $3x+2x+(-x)+(-7x) = 3x+2x-x-7x = -3x$

Each symbol must be dealt with individually.

For the '$a$' terms:     $+4a-2a = 2a$
For the '$b$' terms:     $+3b-5b = -2b$
For the '$c$' terms:     $+c+6c = 7c$

Thus $4a+3b+c+(-2a)+(-5b)+6c = 4a+3b+c-2a-5b+6c$
$$= 2a-2b+7c$$

The algebraic expressions may be tabulated as shown below, forming columns for the $a$'s, $b$'s, $c$'s and $d$'s. Thus:

$$
\begin{array}{l}
+\ 5a - 2b \\
+\ 2a \qquad +\ c \\
\qquad +\ 4b \qquad\quad -\ 5d \\
\underline{-\ \ a +\ \ b - 4c + 3d} \\
\end{array}
$$

Adding gives:     $6a + 3b - 3c - 2d$

$$
\begin{array}{l}
x - 2y + 5z \\
\underline{2x + 3y - 4z} \\
\end{array}
$$

Subtracting gives:     $-x - 5y + 9z$

(Note that $+5z--4z = +5z+4z = 9z$)

48

An alternative method of subtracting algebraic expressions is to 'change the signs of the bottom line and add'. Hence:

$$x - 2y + 5z$$
$$- \ 2x - 3y + 4z$$

Adding gives:  $\quad -x - 5y + 9z$

---

Each term in the first expression is multiplied by $a$, then each term in the first expression is multiplied by $b$, and the two results are added. The usual layout is shown below.

$$2a \ + \ 3b$$
$$a \ + \ b$$

Multiplying by $a \rightarrow \quad 2a^2 + 3ab$
Multiplying by $b \rightarrow \quad \quad + \ 2ab \ + \ 3b^2$

Adding gives: $\quad 2a^2 + 5ab + 3b^2$

---

$$3x \ - \ 2y^2 \ + \ 4xy$$
$$2x \ - \ 5y$$

Multiplying by $2x \rightarrow \quad 6x^2 - 24xy^2 + 8x^2y$
Multiplying by $-5y \rightarrow \quad \quad - \ 20xy^2 \quad \quad - 15xy + 10y^3$

Adding gives: $\quad 6x^2 - 24xy^2 + 8x^2y - 15xy + 10y^3$

---

$2p \div 8pq$ means $\dfrac{2p}{8pq}$ . This can be reduced by cancelling as in arithmetic.

Thus $\dfrac{2p}{8pq} = \dfrac{^1 \not{2} \times \not{p}^{\ 1}}{_4 \not{8} \times \not{p}_1 \times q} = \dfrac{1}{4q}$

*Problem 10* Divide $2x^2+x-3$ by $x-1$

$2x^2+x-3$ is called the dividend and $x-1$ the divisor. The usual layout is shown below with the dividend and divisor both arranged in descending powers of the symbols.

$$
\begin{array}{r}
2x + 3 \\
x-1 \overline{\smash{\big)}\ 2x^2 + x - 3} \\
\underline{2x^2 - 2x} \\
3x - 3 \\
\underline{3x - 3} \\
\cdot \quad \cdot
\end{array}
$$

Dividing the first term of the dividend by the first term of the divisor, i.e.

$$\frac{2x^2}{x}$$

gives $2x$, which is put above the first term of the dividend as shown. The divisor is then multiplied by $2x$. i.e. $2x(x-1) = 2x^2-2x$, which is placed under the dividend as shown. Subtracting gives $3x-3$. The process is then repeated, i.e. the first term of the divisor is divided into $3x$, giving 3, which is placed above the dividend as shown. Then $3(x-1) = 3x-3$, which is placed under the $3x-3$. The remainder, on subtraction, is zero, which completes the process.

**Thus** $(2x^2+x-3) \div (x-1) = (2x+3)$

(A check can be made on this answer by multiplying $(2x+3)$ by $(x-1)$, which should equal $2x^2+x-3$.)

*Problem 11* Simplify $\dfrac{x^3+y^3}{x+y}$

$$
\begin{array}{r}
\textcircled{1} \quad \textcircled{4} \quad \textcircled{7} \\
x^2 - xy + y^2 \\
x+y \overline{\smash{\big)}\ x^3 + 0 + 0 + y^3} \\
\underline{x^3 + x^2y} \\
-x^2y + y^3 \\
-x^2y - xy^2 \\
xy^2 + y^3 \\
\underline{xy^2 + y^3} \\
\cdot \quad \cdot
\end{array}
$$

① $x$ into $x^3$ goes $x^2$. Put $x^2$ above $x^3$.

② $x^2(x+y) = x^3+x^2y$.

③ Subtract.

④ $x$ into $-x^2y$ goes $-xy$. Put $-xy$ above dividend.

⑤ $-xy(x+y) = -x^2y-xy^2$.

⑥ Subtract.

⑦ $x$ into $xy^2$ goes $y^2$. Put $y^2$ above dividend.

⑧ $y^2(x+y) = xy^2+y^3$.

⑨ Subtract.

**Thus** $\dfrac{x^3+y^3}{x+y} = x^2-xy+y^2$

The zero's shown in the dividend are not normally shown, but are included to clarify the subtraction process and to keep similar terms in their respective columns.

$$
\begin{array}{r}
2a^2 - 2ab \; - \;\;\; b^2 \\
2a-b \overline{\smash{\big)}\; 4a^3 - 6a^2b \qquad\quad + 5b^3} \\
\underline{4a^3 - 2a^2b} \\
-4a^2b \qquad\quad + 5b^3 \\
\underline{-4a^2b + 2ab^2} \\
-2ab^2 + 5b^3 \\
\underline{-2ab^2 + \; b^3} \\
4b^3 \\
\end{array}
$$

Thus $\dfrac{4a^3 - 6a^2b + 5b^3}{2a-b} = 2a^2 - 2ab - b^2$, **remainder $4b^3$**

Alternatively, the answer may be expressed as $2a^2 - 2ab - b^2 + \dfrac{4b^3}{2a-b}$

*Further problems on basic operations may be found in section C, problems 1 to 14, page 61.*

**(b) LAWS OF INDICES**

*Problem 13*  Simplify $a^3 b^2 c \times ab^3 c^5$

Grouping like terms gives:  $a^3 \times a \times b^2 \times b^3 \times c \times c^5$
Using the first law of indices gives:  $a^{3+1} \times b^{2+3} \times c^{1+5}$

      i.e.  $a^4 \times b^5 \times c^6$;  i.e.  $a^4 b^5 c^6$

*Problem 14*  Simplify $a^{\frac{1}{2}} b^2 c^{-2} \times a^{\frac{1}{6}} b^{\frac{1}{2}} c$

Using the first law of indices, $a^{\frac{1}{2}} b^2 c^{-2} \times a^{\frac{1}{6}} b^{\frac{1}{2}} c = a^{\frac{1}{2}+\frac{1}{6}} \times b^{2+\frac{1}{2}} \times c^{-2+1}$
$$= a^{\frac{2}{3}} b^{\frac{5}{2}} c^{-1}$$

*Problem 15*  Simplify $\dfrac{a^3 b^2 c^4}{a \; b \; c^{-2}}$ and evaluate when $a = 3, b = \dfrac{1}{8}$ and $c = 2$.

Using the second law of indices, $\dfrac{a^3}{a} = a^{3-1} = a^2, \dfrac{b^2}{b} = b^{2-1} = b$

and $\dfrac{c^4}{c^{-2}} = c^{4--2} = c^6$;  Thus $\dfrac{a^3 b^2 c^4}{a \; b \; c^{-2}} = a^2 b c^6$

When $a = 3, b = \dfrac{1}{8}$ and $c = 2, a^2 b c^6 = (3)^2 \dfrac{1}{8} (2)^6 = (9) \dfrac{1}{8} (64) = 72$

**Problem 16** Simplify $\dfrac{p^{\frac{1}{2}}q^2r^{\frac{2}{3}}}{p^{\frac{1}{4}}q^{\frac{1}{2}}r^{\frac{1}{6}}}$ and evaluate when $p = 16$, $q = 9$ and $r = 4$, taking positive of roots only

Using the second law of indices gives: $\quad p^{\frac{1}{2}-\frac{1}{4}} \; q^{2-\frac{1}{2}} \; r^{\frac{2}{3}-\frac{1}{6}} \; = \; p^{\frac{1}{4}}q^{\frac{3}{2}}r^{\frac{1}{2}}$

When $p = 16$, $q = 9$ and $r = 4$, $p^{\frac{1}{4}}q^{\frac{3}{2}}r^{\frac{1}{2}} = (16)^{\frac{1}{4}}(9)^{\frac{3}{2}}(4)^{\frac{1}{2}}$

$$= (\sqrt[4]{16})\,(\sqrt{9^3})\,(\sqrt{4}) = (2)\,(3^3)\,(2) = \mathbf{108}$$

**Problem 17** Simplify $\dfrac{x^2y^3+xy^2}{xy}$

Algebraic expressions of the form $\dfrac{a+b}{c}$ can be split into $\dfrac{a}{c} + \dfrac{b}{c}$

Thus $\dfrac{x^2y^3+xy^2}{xy} = \dfrac{x^2y^3}{xy} + \dfrac{xy^2}{xy} = x^{2-1}y^{3-1}+x^{1-1}y^{2-1}$

$$= xy^2 + y \quad \text{(since } x^0 = 1\text{, from the sixth law of indices)}$$

**Problem 18** Simplify $\dfrac{x^2y}{xy^2-xy}$

The highest common factor (HCF) of each of the three terms comprising the numerator and denominator is $xy$. Dividing each term by $xy$ gives:

$$\frac{x^2y}{xy^2-xy} = \frac{\dfrac{x^2y}{xy}}{\dfrac{xy^2}{xy} - \dfrac{xy}{xy}} = \frac{x}{y-1}$$

**Problem 19** Simplify $\dfrac{a^2b}{ab^2-a^{\frac{1}{2}}b^3}$

The HCF of each of the three terms is $a^{\frac{1}{2}}b$. Dividing each term by $a^{\frac{1}{2}}b$

gives: $\quad \dfrac{a^2b}{ab^2-a^{\frac{1}{2}}b^3} = \dfrac{\dfrac{a^2b}{a^{\frac{1}{2}}b}}{\dfrac{ab^2}{a^{\frac{1}{2}}b} - \dfrac{a^{\frac{1}{2}}b^3}{a^{\frac{1}{2}}b}} = \dfrac{a^{\frac{3}{2}}}{a^{\frac{1}{2}}b-b^2}$

**Problem 20** Simplify $(p^3)^{\frac{1}{2}}(q^2)^4$

Using the third law of indices gives: $\quad p^{3 \times \frac{1}{2}} q^{2 \times 4}$

i.e. $\quad p^{\frac{3}{2}} q^8$

**Problem 21** Simplify $\dfrac{(mn^2)^3}{(m^{\frac{1}{2}}n^{\frac{1}{4}})^4}$

The brackets indicate that each letter in the bracket must be raised to the power outside.

Using the third law of indices gives: $\dfrac{(mn^2)^3}{(m^{\frac{1}{2}}n^{\frac{1}{4}})^4} = \dfrac{m^{1 \times 3}n^{2 \times 3}}{m^{\frac{1}{2} \times 4}n^{\frac{1}{4} \times 4}} = \dfrac{m^3 n^6}{m^2 n^1}$

Using the second law of indices gives: $\dfrac{m^3 n^6}{m^2 n^1} = m^{3-2}n^{6-1} = mn^5$

**Problem 22** Simplify $(a^3\sqrt{b}\sqrt{c^5})(\sqrt{a}\sqrt[3]{b^2}c^3)$ and evaluate when $a = \dfrac{1}{4}$, $b = 64$ and $c = 1$.

Using the fourth law of indices, the expression can be written as:

$(a^3 b^{\frac{1}{2}} c^{\frac{5}{2}})(a^{\frac{1}{2}} b^{\frac{2}{3}} c^3)$

Using the first law of indices gives: $\quad a^{3 + \frac{1}{2}} b^{\frac{1}{2} + \frac{2}{3}} c^{\frac{5}{2} + 3} = a^{\frac{7}{2}} b^{\frac{7}{6}} c^{\frac{11}{2}}$

It is usual to express the answer in the same form as the question. Hence

$a^{\frac{7}{2}} b^{\frac{7}{6}} c^{\frac{11}{2}} = \sqrt{a^7} \sqrt[6]{b^7} \sqrt{c^{11}}$

When $a = \dfrac{1}{4}, b = 64$ and $c = 1, \sqrt{a^7} \sqrt[6]{b^7} \sqrt{c^{11}} = \sqrt{\left(\dfrac{1}{4}\right)^7} (\sqrt[6]{64^7})\sqrt{1^{11}}$

$= \left(\dfrac{1}{2}\right)^7 (2)^7 (1) = 1$

**Problem 23** Simplify $(a^3 b)(a^{-4}b^{-2})$, expressing the answer with positive indices only.

Using the first law of indices gives: $\quad a^{3 + -4} b^{1 + -2} = a^{-1}b^{-1}$

Using the fifth law of indices gives: $\quad a^{-1}b^{-1} = \dfrac{1}{a^{+1}b^{+1}} = \dfrac{1}{ab}$

**Problem 24** Simplify $\dfrac{d^2\,e^2\,f^{\frac{1}{2}}}{\left(d^{\frac{3}{2}}\,e\,f^{\frac{5}{2}}\right)^2}$ , expressing the answer with positive indices only

Using the third law of indices gives: $\dfrac{d^2\,e^2\,f^{\frac{1}{2}}}{d^{\frac{3}{2}\times 2}\,e^{1\times 2}\,f^{\frac{5}{2}\times 2}} = \dfrac{d^2\,e^2\,f^{\frac{1}{2}}}{d^3\,e^2\,f^5}$

Using the second law of indices gives: $\quad d^{2-3}\,e^{2-2}\,f^{\frac{1}{2}-5}$

$$= d^{-1}\,e^0\,f^{-\frac{9}{2}}$$

$$= d^{-1}\,f^{-\frac{9}{2}} \text{ , since } e^0 = 1 \text{ from the sixth law of indices,}$$

$$= \frac{1}{d\,f^{\frac{9}{2}}} \text{ , from the fifth law of indices.}$$

**Problem 25** Simplify $\dfrac{(x^2 y^{\frac{1}{2}})(\sqrt{x}\ \sqrt[3]{y^2})}{(x^5 y^3)^{\frac{1}{2}}}$

Using the third and fourth laws of indices gives: $\quad \dfrac{(x^2 y^{\frac{1}{2}})(x^{\frac{1}{2}} y^{\frac{2}{3}})}{x^{\frac{5}{2}}\,y^{\frac{3}{2}}}$

Using the first and second laws of indices gives: $\quad x^{2+\frac{1}{2}-\frac{5}{2}}\,y^{\frac{1}{2}+\frac{2}{3}-\frac{3}{2}}$

$$= x^0\,y^{-\frac{1}{3}} = y^{-\frac{1}{3}} \text{ or } \frac{1}{y^{\frac{1}{3}}}$$

from the fifth and sixth laws of indices.

*Further problems on laws of indices may be found in section C, problems 15–25, page 62.*

(c) BRACKETS AND FACTORISATION

**Problem 26** Remove the brackets and simplify the expression $(3a+b)+2(b+c)-4(c+d)$

Both $b$ and $c$ in the second bracket have to be multiplied by 2, and $c$ and $d$ in the third bracket by $-4$ when the brackets are removed (see para 2(v)). Thus:

$(3a+b)+2(b+c)-4(c+d) = 3a+b+2b+2c-4c-4d$

Collecting similar terms together gives: $3a+3b-2c-4d$

When the brackets are removed, both $2a$ and $-ab$ in the first bracket must be multiplied by $-1$ and both $3b$ and $a$ in the second bracket by $-a$. Thus:

$a^2-(2a-ab)-a(3b+a) = a^2-2a+ab-3ab-a^2$

Collecting similar terms together gives: $-2a-2ab$
Since $-2a$ is a common factor the answer can be expressed as $-2a(1+b)$

Each term in the second bracket has to be multiplied by each term in the first bracket. Thus:

$(a+b)(a-b) = a(a-b)+b(a-b)$
$= a^2-ab+ab-b^2 = a^2-b^2$

| Alternatively | $a + b$ |
| | $a - b$ |
| Multiplying by $a \rightarrow$ | $a^2 + ab$ |
| Multiplying by $-b \rightarrow$ | $- ab - b^2$ |
| Adding gives: | $a^2 \qquad - b^2$ |

$(x-2y)(3x+y^2) = x(3x+y^2)-2y(3x+y^2) = 3x^2+xy^2-6xy-2y^3$

$(2x-3y)^2 = (2x-3y)(2x-3y) = 2x(2x-3y)-3y(2x-3y)$
$= 4x^2-6xy-6xy+9y^2$
$= 4x^2-12xy+9y^2$

| Alternatively | $2x - 3y$ |
| | $2x - 3y$ |
| Multiplying by $2x \rightarrow$ | $4x^2- 6xy$ |
| Multiplying by $-3y \rightarrow$ | $- 6xy + 9y^2$ |
| Adding gives: | $4x^2-12xy + 9y^2$ |

*Problem 31* Remove the brackets from the expression $2[p^2-3(q+r)+q^2]$

In this problem there are two brackets and the 'inner' one is removed first.

Hence
$$2[p^2-3(q+r)+q^2] = 2[p^2-3q-3r+q^2]$$
$$= 2p^2-6q-6r+2q^2$$

*Problem 32* Remove the brackets and simplify the expression
$2a-[3\{2(4a-b)-5(a+2b)\}+4a]$

Removing the innermost brackets gives:        $2a-[3\{8a-2b-5a-10b\}+4a]$
Collecting together similar terms gives:        $2a-[3\{3a-12b\}+4a]$
Removing the 'curly' brackets gives:        $2a-[9a-36b+4a]$
Collecting together similar terms gives:        $2a-[13a-36b]$
Removing the outer brackets gives:        $2a-13a+36b$
i.e.        $-11a+36b$ or $36b-11a$ (see para 2(iii))

*Problem 33* Simplify $x(2x-4y)-2x(4x+y)$

Removing brackets gives:        $2x^2-4xy-8x^2-2xy$
Collecting together similar terms gives:        $-6x^2-6xy$
Factorising gives:        $-6x(x+y)$, since $-6x$ is common to
both terms.

*Problem 34* Factorise (a) $xy-3xz$, (b) $4a^2+16ab^3$, (c) $3a^2b-6ab^2+15ab$

For each part of this problem, the HCF of the terms will become one of the
factors.

Thus:    (a)  $xy-3xz$        $= x(y-3z)$
    (b)  $4a^2+16ab^3$        $= 4a(a+4b^3)$
    (c)  $3a^2b-6ab^2+15ab$  $= 3ab(a-2b+5)$

*Problem 35* Factorise $ax-ay+bx-by$

The first two terms have a common factor of $a$ and the last two terms a common
factor of $b$.

Thus:    $ax-ay+bx-by = a(x-y)\quad .-y)$
The two newly formed terms have a common factor of $(x-y)$
Thus:    $a(x-y)+b(x-y) = (x-y)(a+b)$

$a$ is a common factor of the first two terms and $b$ a common factor of the last two terms.

Thus:  $2ax-3ay+2bx-3by = a(2x-3y)+b(2x-3y)$

$(2x-3y)$ is now a common factor thus:

$a(2x-3y)+b(2x-3y) = \mathbf{(2x-3y)(a+b)}$

Alternatively, $2x$ is a common factor of the original first and third terms and $-3y$ is a common factor of the second and fourth terms. Thus:

$2ax-3ay+2bx-3by = 2x(a+b)-3y(a+b)$

$(a+b)$ is now a common factor thus: $2x(a+b)-3y(a+b) = \mathbf{(a+b)(2x-3y)}$, as before (see para 2(iv))

$x^2$ is a common factor of the first two terms thus:

$x^3+3x^2-x-3 = x^2(x+3)-x-3$

$-1$ is a common factor of the last two terms thus:

$x^2(x+3)-x-3 = x^2(x+3)-1(x+3)$

$(x+3)$ is now a common factor thus $x^2(x+3)-1(x+3) = \mathbf{(x+3)(x^2-1)}$

*Further problems on brackets and factorisation may be found in section C, problems 26–42, page 62.*

### (d) FUNDAMENTAL LAWS AND PRECEDENCE

Multiplication is performed before addition and subtraction thus:

$2a+5a \times 3a-a = 2a+15a^2-a$
$= a+15a^2 = \mathbf{a(1+15a)}$

The order of precedence is brackets, multiplication, then subtraction. Hence:

$(a+5a) \times 2a-3a = 6a \times 2a-3a = 12a^2-3a = \mathbf{3a(4a-1)}$

The order of precedence is brackets, multiplication, then subtraction. Hence:

$$a+5a \times (2a-3a) = a+5a \times -a = a+-5a^2$$
$$= a-5a^2 = a(1-5a)$$

The order of precedence is division, then addition and subtraction. Hence:

$$a \div 5a+2a-3a = \frac{a}{5a} + 2a-3a = \frac{1}{5} + 2a-3a = \frac{1}{5} - a$$

The order of precedence is brackets, division and subtraction. Hence:

$$a \div (5a+2a)-3a = a \div 7a-3a = \frac{a}{7a} -3a = \frac{1}{7} -3a$$

The order of precedence is brackets, then division. Hence:

$$a \div (5a+2a-3a) = a \div 4a = \frac{a}{4a} = \frac{1}{4}$$

The order of precedence is division, multiplication, addition and subtraction. Hence:

$$3c+2c \times 4c+c \div 5c-8c = 3c+2c \times 4c+(c/5c)-8c$$

$$= 3c+8c^2 + \frac{1}{5} - 8c = 8c^2-5c+\frac{1}{5} \text{ or } c(8c-5)+\frac{1}{5}$$

The order of precedence is brackets, division, multiplication, addition and subtraction. Hence:

$$(3c+2c)4c+c \div 5c-8c = 5c \times 4c+c \div 5c-8c = 5c \times 4c+ \frac{c}{5c} -8c$$

$$= 20c^2+\frac{1}{5} -8c \text{ or } 4c(5c-2)+\frac{1}{5}$$

The order of precedence is brackets, division, multiplication and addition. Hence:

$3c+2c \times 4c+c \div (5c-8c) = 3c+2c \times 4c+c \div -3c$

$$= 3c+2c \times 4c+\frac{c}{-3c}$$

Now $\frac{c}{-3c} = \frac{1}{-3}$. Multiplying numerator and denominator by $-1$ gives

$\frac{1 \times -1}{-3 \times -1}$ i.e. $-\frac{1}{3}$. Hence:

$3c+2c \times 4c+\frac{c}{-3c} = 3c+2c \times 4c-\frac{1}{3} = 3c+8c^2-\frac{1}{3}$ or $c(3+8c)-\frac{1}{3}$

The order of precedence is brackets, division and multiplication. Hence:

$(3c+2c)(4c+c) \div (5c-8c) = 5c \times 5c \div -3c = 5c \times \frac{5c}{-3c}$

$$= 5c \times -\frac{5}{3} = -\frac{25}{3} c$$

The bracket around the $(2a-3)$ shows that both $2a$ and $-3$ have to be divided by $4a$, and to remove the bracket the expression is written in fraction form. Hence:

$(2a-3) \div 4a+5 \times 6-3a = \frac{2a-3}{4a} +5 \times 6-3a = \frac{2a-3}{4a} + 30-3a$

$$= \frac{2a}{4a} - \frac{3}{4a} + 30-3a = \frac{1}{2} - \frac{3}{4a}+30-3a$$

$$= 30\frac{1}{2}- \frac{3}{4a} - 3a$$

Applying BODMAS, the expression becomes $\frac{1}{3}$ of $3p+4p \times 2p$,

and changing 'of' to '×', gives: $\qquad \frac{1}{3} \times 3p+4p \times 2p$

$\qquad\qquad\qquad$ i.e. $\qquad p+8p^2$ or $p(1+8p)$

*Further problems on fundamental laws and precedence may be found in section C following problems 43 to 54, page 63.*

*Problem 50* If $y$ is directly proportional to $x$ and $y = 2.48$ when $x = 0.4$, determine (a) the coefficient of proportionality and (b) the value of $y$ when $x = 0.65$

(a) $y \propto x$, i.e. $y = kx$

When $y = 2.48$ and $x = 0.4$, $2.48 = k(0.4)$

Hence the coefficient of proportionality, $k = \dfrac{2.48}{0.4} = \textbf{6.2}$

(b) $y = kx$. Hence, when $x = 0.65$, $y = (6.2)(0.65) = \textbf{4.03}$

*Problem 51* Hooke's law states that stress $\sigma$ is directly proportional to strain $\epsilon$ within the elastic limit of a material. When, for mild steel, the stress is $25 \times 10^6$ pascals, the strain is 0.000 125. Determine (a) the coefficient of proportionality and (b) the value of strain when the stress is $18 \times 10^6$ pascals.

(a) $\sigma \propto \epsilon$, i.e. $\sigma = k\epsilon$, from which $k = \dfrac{\sigma}{\epsilon}$

Hence the coefficient of proportionality, $k = \dfrac{25 \times 10^6}{0.000\ 125} = \textbf{200} \times \textbf{10}^{\textbf{9}}$ **pascals**

(The coefficient of proportionality $k$ in this case is called Young's Modulus of Elasticity.)

(b) Since $\sigma = k\epsilon$, $\epsilon = \dfrac{\sigma}{k}$

Hence when $\sigma = 18 \times 10^6$, strain $\epsilon = \dfrac{18 \times 10^6}{200 \times 10^9} = \textbf{0.000\ 09}$

*Problem 52* The electrical resistance $R$ of a piece of wire is inversely proportional to the cross-sectional area $A$. When $A = 5$ mm$^2$, $R = 7.2$ ohms. Determine (a) the coefficient of proportionality and (b) the cross-sectional area when the resistance is 4 ohms.

(a) $R \propto 1/A$, i.e. $R = k/A$ or $k = RA$

Hence, when $R = 7.2$ and $A = 5$, the coefficient of proportionality, $k = (7.2)(5) = \textbf{36}$

(b) Since $k = RA$ then $A = \dfrac{k}{R}$

When $R = 4$, the cross sectional area, $A = \dfrac{36}{4} = \textbf{9 mm}^{\textbf{2}}$

Problem 53 Boyle's law states that at constant temperature, the volume $V$ of a fixed mass of gas is inversely proportional to its absolute pressure $p$. If a gas occupies a volume of 0.08 m³ at a pressure of $1.5 \times 10^6$ pascals determine (a) the coefficient of proportionality and (b) the volume if the pressure is changed to $4 \times 10^6$ pascals.

(a) $V \propto \dfrac{1}{p}$, i.e. $V = \dfrac{k}{p}$ or $k = pV$.

  Hence the coefficient of proportionality, $k = (1.5 \times 10^6)(0.08) = \mathbf{0.12 \times 10^6}$

(b) Volume $V = \dfrac{k}{p} = \dfrac{0.12 \times 10^6}{4 \times 10^6} = \mathbf{0.03 \text{ m}^3}$

*Further problems on direct and inverse proportionality may be found in section C following, Problems 55 to 59, page 64.*

## C. FURTHER PROBLEMS ON ALGEBRA

*Basic operations*

1  Find the value of $2xy + 3yz - xyz$, when $x = 2$, $y = -2$ and $z = 4$

$$[-16]$$

2  Evaluate $3pq^2r^3$ when $p = \dfrac{2}{3}$, $q = -2$ and $r = -1$

$$[-8]$$

3  Find the sum of $3a$, $-2a$, $-6a$, $5a$ and $4a$

$$[4a]$$

4  Simplify $\dfrac{4}{3}c + 2c - \dfrac{1}{6}c - c$

$$\left[2\dfrac{1}{6}c\right]$$

5  Find the sum of $3x$, $2y$, $-5x$, $2z$, $-\dfrac{1}{2}y$, $-\dfrac{1}{4}x$

$$\left[-2\dfrac{1}{4}x + 1\dfrac{1}{2}y + 2z\right]$$

6  Add together $2a + 3b + 4c$, $-5a - 2b + c$, $4a - 5b - 6c$

$$[a - 4b - c]$$

7  Add together $3d + 4e$, $-2e + f$, $2d - 3f$, $4d - e + 2f - 3e$

$$[9d - 2e]$$

8  From $4x - 3y + 2z$ subtract $x + 2y - 3z$

$$[3x - 5y + 5z]$$

9  Subtract $\dfrac{3}{2}a - \dfrac{b}{3} + c$ from $\dfrac{b}{2} - 4a - 3c$

$$\left[-5\dfrac{1}{2}a + \dfrac{5}{6}b - 4c\right]$$

10  Multiply $3x + 2y$ by $x - y$

$$[3x^2 - xy - 2y^2]$$

11  Multiply $2a - 5b + c$ by $3a + b$

$$[6a^2 - 13ab + 3ac - 5b^2 + bc]$$

12  Simplify (i) $3a \div 9ab$; (ii) $4a^2b \div 2a$

$$\left[\text{(i) } \dfrac{1}{3b}; \text{ (ii) } 2ab\right]$$

13  Divide $2x^2 + xy - y^2$ by $x + y$

$$[2x - y]$$

14  Divide $p^3 + q^3$ by $p + q$

$$[p^2 - pq + q^2]$$

*Laws of indices*

15  Simplify $(x^2 y^3 z)(x^3 yz^2)$ and evaluate when $x = \frac{1}{2}$, $y = 2$ and $z = 3$.

$$\left[ x^5 y^4 z^3 \, ; \, 13\tfrac{1}{2} \right]$$

16  Simplify $(a^{\frac{3}{2}} bc^{-3})(a^{\frac{1}{2}} b^{-\frac{1}{2}} c)$ and evaluate when $a = 3$, $b = 4$ and $c = 2$.

$$\left[ a^2 b^{\frac{1}{2}} c^{-2} \, ; \, 4\tfrac{1}{2} \right]$$

17  Simplify $\dfrac{a^5 bc^3}{a^2 b^3 c^2}$ and evaluate when $a = \frac{3}{2}$, $b = \frac{1}{2}$ and $c = \frac{2}{3}$.

$$[a^3 b^{-2} c \, ; \, 9]$$

In *Problems 18 to 25*, simplify the given expressions

18  $\dfrac{x^{\frac{1}{5}} y^{\frac{1}{2}} z^{\frac{1}{3}}}{x^{-\frac{1}{2}} y^{\frac{1}{3}} z^{-\frac{1}{6}}}$

$$\left[ x^{\frac{7}{10}} y^{\frac{1}{6}} z^{\frac{1}{2}} \right]$$

19  $\dfrac{a^2 b + a^3 b}{a^2 b^2}$

$$\left[ \dfrac{1+a}{b} \right]$$

20  $\dfrac{p^3 q^2}{pq^2 - p^2 q}$

$$\left[ \dfrac{p^2 q}{q-p} \right]$$

21  $(a^2)^{\frac{1}{2}} (b^2)^3 (c^{\frac{1}{2}})^3$

$$\left[ ab^6 c^{\frac{3}{2}} \right]$$

22  $\dfrac{(abc)^2}{(a^2 b^{-1} c^{-3})^3}$

$$[a^{-4} b^5 c^{11}]$$

23  $(\sqrt{x} \sqrt{y^3} \sqrt[3]{z^2})(\sqrt{x} \sqrt{y^3} \sqrt{z^3})$

$$[xy^3 \sqrt[6]{z^{13}}]$$

24  $(e^2 f^3)(e^{-3} f^{-5})$, expressing the answer with positive indices only.

$$\left[ \dfrac{1}{ef^2} \right]$$

25  $\dfrac{(a^3 b^{\frac{1}{2}} c^{-\frac{1}{2}})(ab)^{\frac{1}{3}}}{(\sqrt{a^3} \sqrt{b} \, c)}$

$$\left[ a^{\frac{11}{6}} b^{\frac{1}{3}} c^{-\frac{3}{2}} \text{ or } \dfrac{\sqrt[6]{a^{11}} \sqrt[3]{b}}{\sqrt{c^3}} \right]$$

*Brackets and factorisation*

In *Problems 26 to 38*, remove the brackets and simplify where possible.

26  $(x+2y)+(2x-y)$  $\hfill [3x+y]$

27  $(4a+3y)-(a-2y)$

$$[3a+5y]$$

28  $2(x-y)-3(y-x)$

$$[5(x-y)]$$

29  $2x^2 - 3(x-xy) - x(2y-x)$

$$[x(3x-3+y)]$$

30  $2(p+3q-r)-4(r-q+2p)+p$

$$[-5p+10q-6r]$$

31  $(a+b)(a+2b)$

$$[a^2+3ab+2b^2]$$

32  $(p+q)(3p-2q)$

$$[3p^2+pq-2q^2]$$

62

33  (i) $(x-2y)^2$,  (ii) $(3a-b)^2$

$$\begin{bmatrix} \text{(i) } x^2-4xy+4y^2 \\ \text{(ii) } 9a^2-6ab+b^2 \end{bmatrix}$$

34  $3a(b+c)+4c(a-b)$

$$[3ab+7ac-4bc]$$

35  $2x+\{y-(2x+y)\}$

$$[0]$$

36  $3a+2\{a-(3a-2)\}$

$$[4-a]$$

37  $2-5\{a(a-2b)-(a-b)^2\}$

$$[2+5b^2]$$

38  $24p-[2\{3(5p-q)-2(p+2q)\}+3q]$

$$[11q-2p]$$

In *Problems 39 to 42*, factorise.

39  (i) $pb+2pc$,  (ii) $2l^2+8ln$

$$\begin{bmatrix} \text{(i) } p(b+2c) \\ \text{(ii) } 2l(l+4n) \end{bmatrix}$$

40  (i) $21a^2b^2-28ab$, (ii) $2xy^2+6x^2y+8x^3y$

$$\begin{bmatrix} \text{(i) } 7ab(3ab-4) \\ \text{(ii) } 2xy(y+3x+4x^2) \end{bmatrix}$$

41  (i) $ay+by+a+b$, (ii) $px+qx+py+qy$

$$\begin{bmatrix} \text{(i) } (a+b)(y+1) \\ \text{(ii) } (p+q)(x+y) \end{bmatrix}$$

42  (i) $ax-ay+bx-by$, (ii) $2ax+3ay-4bx-6by$

$$\begin{bmatrix} \text{(i) } (x-y)(a+b) \\ \text{(ii) } (a-2b)(2x+3y) \end{bmatrix}$$

*Fundamental laws and precedence*
In *Problems 43 to 54*, simplify.

43  $2x \div 4x+6x$

$$\left[\frac{1}{2}+6x\right]$$

44  $2x \div (4x+6x)$

$$\left[\frac{1}{5}\right]$$

45  $3a-2a \times 4a+a$

$$[4a(1-2a)]$$

46  $(3a-2a)4a+a$

$$[a(4a+1)]$$

47  $3a-2a(4a+a)$

$$[a(3-10a)]$$

48  $2y+4 \div 6y+3 \times 4-6y$

$$\left[\frac{2}{3y}-3y+12\right]$$

49  $(2y+4) \div 6y+3 \times 4-5y$

$$\left[\frac{2}{3y}+12\frac{1}{3}-5y\right]$$

50  $2y+4 \div 6y+3(4-5y)$

$$\left[\frac{2}{3y}+12-13y\right]$$

51  $3 \div y+2 \div y+1$

$$\left[\frac{5}{y}+1\right]$$

52  $p^2-3pq \times 2p \div 6q+pq$

$$[pq]$$

63

53  $(x+1)(x-4) \div (2x+2)$

54  $\frac{1}{4}$of $2y+3y(2y-y)$

$\left[\frac{1}{2}(x-4)\right]$

$\left[y(\frac{1}{2}+3y)\right]$

*Direct and inverse proportionality*

55  If $p$ is directly proportional to $q$ and $p = 37.5$ when $q = 2.5$, determine (a) the constant of proportionality and (b) the value of $p$ when $q$ is 5.2.

[(a) 15; (b) 78]

56  Charle's law states that for a given mass of gas at constant pressure the volume is directly proportional to its thermodynamic temperature. A gas occupies a volume of 2.25 litres at 300 K. Determine (a) the constant of proportionality, (b) the volume at 420 K and (c) the temperature when the volume is 2.625 litres.

[(a) 0.0075; (b) 3.15 l; (c) 350 K]

57  Ohm's law states that the current flowing in a fixed resistor is directly proportional to the applied voltage. When 30 volts is applied across a resistor the current flowing through the resistor is $2.4 \times 10^{-3}$ amperes. Determine (a) the constant of proportionality, (b) the current when the voltage is 52 volts, and (c) the voltage when the current is $3.6 \times 10^{-3}$ amperes.

[(a) 0.000 08; (b) $4.16 \times 10^{-3}$ A; (c) 45 V]

58  If $y$ is inversely proportional to $x$ and $y = 15.3$ when $x = 0.6$, determine (a) the coefficient of proportionality, (b) the value of $y$ when $x$ is 1.5, and (c) the value of $x$ when $y$ is 27.2.

[(a) 9.18; (b) 6.12; (c) 0.337 5]

59  Boyle's law states that for a gas at constant temperature, the volume of a fixed mass of gas is inversely proportional to its absolute pressure. If a gas occupies a volume of 1.5 m³ at a pressure of $200 \times 10^3$ pascals, determine (a) the constant of proportionality, (b) the volume when the pressure is $800 \times 10^3$ pascals and (c) the pressure when the volume is 1.25 m³.

[(a) $300 \times 10^3$ ; (b) 0.375 m³ ; (c) $240 \times 10^3$ Pa]

# 5 Simple equations

## A. MAIN POINTS CONCERNED WITH SIMPLE EQUATIONS

1   $(3x-5)$ is an example of an **algebraic expression**, whereas $3x-5 = 1$ is an example of an **equation** (i.e. it contains an 'equals' sign).
2   An **equation** is simply a statement that two quantities are equal. For example, $1m = 1\,000$ mm or $y = mx+c$.
3   An **identity** is a relationship which is true for all values of the unknown, whereas an equation is only true for particular values of the unknown. For example, $3x-5 = 1$ is an equation since it is only true when $x = 2$, whereas $3x \equiv 8x-5x$ is an identity since it is true for **all** values of $x$. (Note '$\equiv$' means 'is identical to'.)
4   **Simple linear equations** (or **equations of the first degree**) are those in which an unknown quantity is raised only to the power 1.
5   To 'solve an equation' means 'to find the value of the unknown'.
6   Any arithmetic operation may be applied to an equation **as long as the equality of the equation is maintained**.

## B. WORKED PROBLEMS ON SIMPLE EQUATIONS

*Problem 1* Solve the equation $4x = 20$

Dividing each side of the equation by 4 gives: $\dfrac{4x}{4} = \dfrac{20}{4}$

(Note that the same operation has been applied to both the left hand side (LHS) and the right hand side (RHS) of the equation so the equality has been maintained.)
    Cancelling gives $x = 5$, which is the solution to the equation. Solutions to simple equations should always be checked and this is accomplished by substituting the solution into the original equation. In this case, LHS = $4(5) = 20$ = RHS.

*Problem 2* Solve $\dfrac{2x}{5} = 6$

The LHS is a fraction and this can be removed by multiplying both sides of the equation by 5. Hence $5(2x/5) = 5(6)$.
Cancelling gives:   $2x = 30$.

65

Dividing both sides of the equation by 2 gives: $\dfrac{2x}{2} = \dfrac{30}{2}$

i.e. $\quad x = 15$

*Problem 3* Solve $a-5 = 8$

Adding 5 to both sides of the equation gives: $\quad a-5+5 = 8+5$

i.e. $\quad a = 13$

The result of the above procedure is to move the '$-5$' from the LHS of the original equation, across the equals sign, to the RHS, **but the sign is changed to +.**

*Problem 4* Solve $x+3 = 7$

Subtracting 3 from both sides of the equation gives: $x+3-3 = 7-3$

i.e. $\quad x = 4$

The result of the above procedure is to move the '$+3$' from the LHS of the original equation, across the equals sign, to the RHS, **but the sign is changed to $-$.** Thus a term can be moved from one side of an equation to the other as long as a change in sign is made.

*Problem 5* Solve $6x+1 = 2x+9$

In such equations the terms containing $x$ are grouped on one side of the equation and the remaining terms grouped on the other side of the equation. As in problems 3 and 4, changing from one side of an equation to the other must be accompanied by a change of sign.

Thus since $6x+1 \quad = 2x+9$

then $6x-2x = 9-1$

$\qquad 4x = 8$

$$\dfrac{4x}{4} = \dfrac{8}{4}$$

i.e. $\qquad x = 2$

Check: LHS of original equation $= 6(2)+1 = 13$

RHS of original equation $= 2(2)+9 = 13$

Hence the solution $x = 2$ is correct.

*Problem 6* Solve $4-3p = 2p-11$

In order to keep the $p$ term positive the terms in $p$ are moved to the RHS and the constant terms to the LHS.

Hence $\quad 4+11 = 2p+3p$

$\qquad\qquad 15 = 5p$

$\qquad\qquad \dfrac{15}{5} = \dfrac{5p}{5}$

Hence $\qquad p = 3$

Check: $\quad$ LHS $= 4-3(3) = 4-9 = -5$

$\qquad\qquad$ RHS $= 2(3)-11 = 6-11 = -5$

$\qquad\qquad$ Hence the solution $p = 3$ is correct.

If, in this example, the unknown quantities had been grouped initially on the LHS instead of the RHS then:

$\qquad\qquad\qquad -3p-2p = -11-4$

i.e. $\qquad\qquad -5p = -15$

$\qquad\qquad\qquad \dfrac{-5p}{-5} = \dfrac{-15}{-5}$

and $\qquad\qquad p = 3$, as before.

It is often easier, however, to work with positive values where possible.

---

*Problem 7* Solve $3(x-2) = 9$

---

Removing the bracket gives: $\quad 3x-6 = 9$

Rearranging gives: $\qquad\qquad 3x = 9+6$

$\qquad\qquad\qquad\qquad 3x = 15$

$\qquad\qquad\qquad\qquad \dfrac{3x}{3} = \dfrac{15}{3}$

i.e. $\quad x = 5$

Check: $\quad$ LHS $= 3(5-2) = 3(3) = 9 =$ RHS

$\qquad\qquad$ Hence the solution $x = 5$ is correct.

---

*Problem 8* Solve $4(2r-3)-2(r-4) = 3(r-3)-1$

---

Removing brackets gives: $\quad 8r-12-2r+8 = 3r-9-1$

Rearranging gives: $\qquad\qquad 8r-2r-3r = -9-1+12-8$

i.e. $\qquad\qquad\qquad 3r = -6$

$\qquad\qquad\qquad\qquad r = \dfrac{-6}{3} = -2$

Check: $\quad$ LHS $= 4(-4-3)-2(-2-4) = -28+12 = -16$

$\qquad\qquad$ RHS $= 3(-2-3)-1 = -15-1 = -16$

$\qquad\qquad$ Hence the solution $r = -2$ is correct.

---

*Problem 9* Solve $\dfrac{3}{x} = \dfrac{4}{5}$

---

The lowest common multiple (LCM) of the denominators, i.e. the lowest algebraic expression that both $x$ and 5 will divide into, is $5x$.

Multiplying both sides by $5x$ gives: $\quad 5x\left(\dfrac{3}{x}\right) = 5x\left(\dfrac{4}{5}\right)$

Cancelling gives: $\qquad\qquad\qquad\qquad 15 = 4x \qquad\qquad\qquad (1)$

$$\frac{15}{4} = \frac{4x}{4}$$

i.e. $\qquad\qquad x = 3\dfrac{3}{4}$

Check: LHS $= \dfrac{3}{3\dfrac{3}{4}} = \dfrac{3}{15/4} = 3\left(\dfrac{4}{15}\right) = \dfrac{12}{15} = \dfrac{4}{5} = $ RHS

(Note that when there is only one fraction on each side of an equation, 'cross-multiplication' can be applied. In this example if $\dfrac{3}{x} = \dfrac{4}{5}$ then $(3)(5) = 4x$, which is a quicker way of arriving at equation (1) above.)

---

***Problem 10*** Solve $\dfrac{2y}{5} + \dfrac{3}{4} + 5 = \dfrac{1}{20} - \dfrac{3y}{2}$

The LCM of the denominators is 20.

Multiplying each term by 20 gives: $20\left(\dfrac{2y}{5}\right) + 20\left(\dfrac{3}{4}\right) + 20(5) = 20\left(\dfrac{1}{20}\right) - 20\left(\dfrac{3y}{2}\right)$

Cancelling gives: $\qquad\qquad 4(2y) + 5(3) + 100 = 1 - 10(3y)$

i.e. $\qquad\qquad\qquad 8y + 15 + 100 = 1 - 30y$

Rearranging gives: $\qquad\qquad 8y + 30y = 1 - 15 - 100$

$$38y = -114$$

$$y = \frac{-114}{38} = -3$$

Check: LHS $= \dfrac{2(-3)}{5} + \dfrac{3}{4} + 5 = \dfrac{-6}{5} + \dfrac{3}{4} + 5 = \dfrac{-9}{20} + 5 = 4\dfrac{11}{20}$

RHS $= \dfrac{1}{20} - \dfrac{3(-3)}{2} = \dfrac{1}{20} + \dfrac{9}{2} = 4\dfrac{11}{20}$

Hence the solution $y = -3$ is correct.

---

***Problem 11*** Solve $\dfrac{3}{t-2} = \dfrac{4}{3t+4}$

By 'cross-multiplication': $\qquad 3(3t+4) = 4(t-2)$

Removing brackets gives: $\qquad 9t + 12 = 4t - 8$

Rearranging gives: $\qquad\qquad 9t - 4t = -8 - 12$

i.e. $\qquad\qquad\qquad\qquad 5t = -20$

$$t = \frac{-20}{5} = -4$$

Check:   LHS $= \dfrac{3}{-4-2} = \dfrac{3}{-6} = -\dfrac{1}{2}$

RHS $= \dfrac{4}{3(-4)+4} = \dfrac{4}{-12+4} = \dfrac{4}{-8} = -\dfrac{1}{2}$

Hence the solution $t = -4$ is correct.

**Problem 12** Solve $\sqrt{x} = 2$

Wherever square root signs are involved with the unknown quantity, both sides of the equation must be squared. Hence $(\sqrt{x})^2 = (2)^2$

i.e.    $x = \mathbf{4}$

**Problem 13** Solve $2\sqrt{d} = 8$

To avoid possible errors it is usually best to arrange the term containing the square root on its own.

Thus $\dfrac{2\sqrt{d}}{2} = \dfrac{8}{2}$

i.e.   $\sqrt{d} = 4$

Squaring both sides gives: $d = \mathbf{16}$, which may be checked in the original equation.

**Problem 14** Solve $\left(\dfrac{\sqrt{b}+3}{\sqrt{b}}\right) = 2$

To remove the fraction each term is multiplied by $\sqrt{b}$.

Hence $\sqrt{b}\left(\dfrac{\sqrt{b}+3}{\sqrt{b}}\right) = \sqrt{b}(2)$

Cancelling gives: $\sqrt{b}+3 = 2\sqrt{b}$
Rearranging gives:    $3 = 2\sqrt{b}-\sqrt{b} = \sqrt{b}$
Squaring both sides gives:    $9 = b$

Check:   LHS $= \dfrac{\sqrt{9}+3}{\sqrt{9}} = \dfrac{3+3}{3} = \dfrac{6}{3} = 2 =$ RHS

**Problem 15** Solve $x^2 = 25$

This problem (and problem 16) involves a square term and thus are not simple equations (they are, in fact, quadratic equations). However the solution of such equations are often required and are therefore included for completeness.

Whenever a square of the unknown is involved, the square root of both sides of the equation is taken. Hence $\sqrt{x^2} = \sqrt{25}$

i.e. $x = 5$

However, $x = -5$ is also a solution of the equation because $(-5) \times (-5) = +25$. Therefore, whenever the square root of a number is required there are always **two** answers, one positive, the other negative.

The solution of $x^2 = 25$ is thus written as $x = \pm 5$

*Problem 16*  Solve $\dfrac{15}{4t^2} = \dfrac{2}{3}$

'Cross-multiplying' gives:  $15(3) = 2(4t^2)$

$$45 = 8t^2$$

$$\frac{45}{8} = t^2$$

i.e.  $t^2 = 5\dfrac{5}{8} = 5.625$

Hence  $t = \sqrt{(5.625)} = \pm 2.372$ correct to 4 significant figures.

*Further problems involving simple equations may be found in section C, problems 1 to 44, page 74.*

## PRACTICAL PROBLEMS INVOLVING SIMPLE EQUATIONS

*Problem 17*  A copper wire has a length $l$ of 1.5 km, a resistance $R$ of 5 $\Omega$ and a resistivity $\rho$ of $17.2 \times 10^{-6}$ $\Omega$ mm. Find the cross-sectional areas, $a$, of the wire, given that $R = \rho l/a$

Since $R = \dfrac{\rho l}{a}$ then $5 \ \Omega = \dfrac{(17.2 \times 10^{-6} \ \Omega \ \text{mm})(1\ 500 \times 10^3 \ \text{mm})}{a}$

From the units given, $a$ is measured in mm$^2$. Thus

$5a = 17.2 \times 10^{-6} \times 1\ 500 \times 10^3$

$a = \dfrac{17.2 \times 10^{-6} \times 1\ 500 \times 10^3}{5}$

$= \dfrac{17.2 \times 1\ 500 \times 10^3}{10^6 \times 5} = \dfrac{17.2 \times \cancel{15}^{\,3}}{10 \times \cancel{5}_{\,1}} = 5.16$

**Hence the cross-sectional area of the wire is 5.16 mm$^2$**

*Problem 18*  A rectangular box with square ends has its length 15 cm greater than its breadth and the total length of its edges if 2.04 m. Find the width of the box and its volume.

Let $x$ cm = width = height of box. Then the length of the box is $(x+15)$ cm
The length of the edges of the box is:  $2(4x)+4(x+15)$ cm
Hence      $204 = 2(4x)+4(x+15)$

$204 = 8x+4x+60$

$204-60 = 12x$

i.e.      $144 = 12x$

and      $x = 12$ cm

**Hence the width of the box is 12 cm**

**Volume of box** = length $\times$ width $\times$ height = $(x+15)(x)(x) = (27)(12)(12)$

$= 3\,888$ cm$^3$

---

*Problem 19* The temperature coefficient of resistance $\alpha$ may be calculated from the formula $R_t = R_0(1+\alpha t)$. Find $\alpha$ given $R_t = 0.928$, $R_0 = 0.8$ and $t = 40$

Since $R_t = R_0(1+\alpha t)$ then $0.928 = 0.8[1+\alpha(40)]$

$0.928 = 0.8+(0.8)(\alpha)(40)$

$0.928-0.8 = 32\alpha$

$0.128 = 32\alpha$

Hence                                        $\alpha = \dfrac{0.128}{32} = \mathbf{0.004}$

---

*Problem 20* The distance $s$ metres travelled in time $t$ seconds is given by the formula $s = ut+\frac{1}{2}at^2$, where $u$ is the initial velocity in m/s and $a$ is the acceleration in m/s$^2$. Find the acceleration of the body if it travels 168 m in 6 s, with an initial velocity of 10 m/s.

$s = ut+\dfrac{1}{2}at^2$, and $s = 168$, $u = 10$ and $t = 6$

Hence      $168 = (10)(6) + \dfrac{1}{2}a(6)^2$

$168 = 60+18a$

$168-60 = 18a$

$108 = 18a$

$a = \dfrac{108}{18} = 6$

**Hence the acceleration of the body is 6 m/s$^2$**

---

*Problem 21* When three resistors in an electrical circuit are connected in parallel the total resistance $R_T$ is given by:

$$\frac{1}{R_T} = \frac{1}{R_1} + \frac{1}{R_2} + \frac{1}{R_3}$$

Find the total resistance when $R_1 = 5\ \Omega$, $R_2 = 10\ \Omega$ and $R_3 = 30\ \Omega$

$$\frac{1}{R_T} = \frac{1}{5} + \frac{1}{10} + \frac{1}{30} = \frac{6+3+1}{30} = \frac{10}{30} = \frac{1}{3}$$

Taking the reciprocal of both sides gives: $R_T = 3\,\Omega$

Alternatively, if $\dfrac{1}{R_T} = \dfrac{1}{5} + \dfrac{1}{10} + \dfrac{1}{30}$ the LCM of the denominators is $30\,R_T$

Hence $\qquad 30\,R_T\left(\dfrac{1}{R_T}\right) = 30\,R_T\left(\dfrac{1}{5}\right) + 30\,R_T\left(\dfrac{1}{10}\right) + 30\,R_T\left(\dfrac{1}{30}\right)$

Cancelling gives: $\quad 30 = 6\,R_T + 3\,R_T + R_T$

i.e. $\qquad 30 = 10\,R_T$

$$R_T = \frac{30}{10} = 3\,\Omega, \text{ as above}$$

---

*Problem 22* The extension $x$ m of an aluminium tie bar of length $l$ m and cross-sectional area $A$ m$^2$ when carrying a load of $F$ newtons is given by the modulus of elasticity $E = Fl/Ax$. Find the extension of the tie bar (in mm) if $E = 70 \times 10^9$ N/m$^2$, $F = 20 \times 10^6$ N, $A = 0.1$ m$^2$ and $l = 1.4$ m.

$E = \dfrac{Fl}{Ax}$ . Hence $70 \times 10^9\ \dfrac{\text{N}}{\text{m}^2} = \dfrac{(20 \times 10^6\ \text{N})(1.4\ \text{m})}{(0.1\ \text{m}^2)(x)}$

(the unit of $x$ is thus metres)

$70 \times 10^9 \times 0.1 \times x = 20 \times 10^6 \times 1.4$

$x = \dfrac{20 \times 10^6 \times 1.4}{70 \times 10^9 \times 0.1}$

Cancelling gives: $\quad x = \dfrac{2 \times 1.4}{7 \times 100}\ \text{m} = \dfrac{2 \times 1.4}{7 \times 100} \times 1000\ \text{mm}$

**Hence the extension of the tie bar, $x = 4$ mm**

---

*Problem 23* Power in a dc circuit is given by $P = V^2/R$, where $V$ is the supply voltage and $R$ is the circuit resistance. Find the supply voltage if the circuit resistance is $1.25\,\Omega$ and the power measured is 320 W

Since $P = \dfrac{V^2}{R}$ then $320 = \dfrac{V^2}{1.25}$

$\qquad (320)(1.25) = V^2$

i.e. $\qquad V^2 = 400$

**Supply voltage,** $\quad V = \sqrt{400} = \pm 20$ **V**

---

*Problem 24* A painter is paid £2.25 per hour for a basic 36 hour week, and overtime is paid at one and a third times this rate. Determine how many hours the painter has to work in a week to earn £114.

Basic rate per hour = £2.25; Overtime rate per hour = $1\frac{1}{3} \times$ £2.25 = £3.

Let the number of overtime hours worked = $x$

Then $(36)(2.25)+(x)(3) = 114$

$$81 + 3x = 114$$
$$3x = 114-81 = 33$$
$$x = \frac{33}{3} = 11$$

Thus 11 hours overtime would have to be worked to earn £114 per week. Hence the total number of hours worked is 36+11, i.e. **47 hours.**

*Problem 25* A formula relating initial and final states of pressures, $P_1$ and $P_2$, volumes, $V_1$ and $V_2$, and absolute temperatures, $T_1$ and $T_2$, of an ideal gas is

$$\frac{P_1 V_1}{T_1} = \frac{P_2 V_2}{T_2}$$

Find the value of $P_2$ given $P_1 = 100 \times 10^3$, $V_1 = 1.0$, $V_2 = 0.266$, $T_1 = 423$ and $T_2 = 293$.

Since $\frac{P_1 V_1}{T_1} = \frac{P_2 V_2}{T_2}$ then $\frac{(100 \times 10^3)(1.0)}{423} = \frac{P_2(0.266)}{293}$

'Cross-multiplying' gives: $(100 \times 10^3)(1.0)(293) = P_2(0.266)(423)$

$$P_2 = \frac{(100 \times 10^3)(1.0)(293)}{(0.266)(423)}$$

**Hence $P_2 = 260 \times 10^3$ or $2.6 \times 10^5$**

*Problem 26* The stress $f$ in a material of a thick cylinder can be obtained from $\frac{D}{d} = \sqrt{\left(\frac{f+p}{f-p}\right)}$. Calculate the stress given that $D = 21.5$, $d = 10.75$ and $p = 1\,800$

Since $\frac{D}{d} = \sqrt{\left(\frac{f+p}{f-p}\right)}$ then $\frac{21.5}{10.75} = \sqrt{\left(\frac{f+1\,800}{f-1\,800}\right)}$

i.e. $2 = \sqrt{\left(\frac{f+1\,800}{f-1\,800}\right)}$

Squaring both sides gives: $4 = \frac{f+1\,800}{f-1\,800}$

$$4(f-1\,800) = f+1\,800$$
$$4f-7\,200 = f+1\,800$$
$$4f-f = 1\,800+7\,200$$
$$3f = 9\,000$$
$$f = \frac{9\,000}{3} = 3\,000$$

**Hence the stress, $f$, is 3000**

*Problem 27* 12 workmen employed on a building site earn between them a total of £815 per week. Labourers are paid £58 per week and craftsmen are paid £75 per week. How many craftsmen and how many labourers are employed?

Let the number of craftsmen be $c$. The number of labourers are thus $(12-c)$. The wage bill equation is:

$$75c+58(12-c) = 815$$
$$75c+696-58c = 815$$
$$75c-58c = 815-696$$
$$17c = 119$$
$$c = \frac{119}{17} = 7$$

**Hence there are 7 craftsmen and $(12-7)$, i.e. 5 labourers on the site**

*Further examples on practical problems involving simple equations may be found in section C, problems 45–58, page 76.*

## C. FURTHER PROBLEMS ON SIMPLE EQUATIONS

Solve the following equations:

1   $2x+5 = 7$ [1]
2   $8-3t = 2$ [2]

3   $\frac{2}{3}c-1 = 3$ [6]

4   $2x-1 = 5x+11$ [−4]

5   $7-4p = 2p-3$ $[1\frac{2}{3}]$

6   $2.6x-1.3 = 0.9x+0.4$ [1]
7   $2a+6-5a = 0$ [2]

8   $3x-2-5x = 2x-4$ $\left[\frac{1}{2}\right]$

9   $20d-3+3d = 11d+5-8$ [0]
10  $2(x-1) = 4$ [3]
11  $16 = 4(t+2)$ [2]
12  $5(f-2)-3(2f+5)+15 = 0$ [−10]
13  $2x = 4(x-3)$ [6]
14  $6(2-3y)-42 = -2(y-1)$ [−2]

15  $2(3g-5)-5 = 0$ $\left[2\frac{1}{2}\right]$

16  $4(3x+1) = 7(x+4)-2(x+5)$ [2]

17  $10+3(r-7) = 16-(r+2)$                                         $\left[6\dfrac{1}{4}\right]$

18  $8+4(x-1)-5(x-3) = 2(5-2x)$                                    $[-3]$

19  $\dfrac{1}{5}d+3 = 4$                                          $[5]$

20  $2+\dfrac{3}{4}y = 1+\dfrac{2}{3}y+\dfrac{5}{6}$              $[-2]$

21  $\dfrac{1}{4}(2x-1)+3 = \dfrac{1}{2}$                         $\left[-4\dfrac{1}{2}\right]$

22  $\dfrac{1}{5}(2f-3)+\dfrac{1}{6}(f-4)+\dfrac{2}{15} = 0$      $[2]$

23  $\dfrac{1}{3}(3m-6)-\dfrac{1}{4}(5m+4)+\dfrac{1}{5}(2m-9) = -3$   $[12]$

24  $\dfrac{x}{3} - \dfrac{x}{5} = 2$                             $[15]$

25  $1-\dfrac{y}{3} = 3+\dfrac{y}{3}-\dfrac{y}{6}$               $[-4]$

26  $\dfrac{2}{a} = \dfrac{3}{8}$                                 $\left[5\dfrac{1}{3}\right]$

27  $\dfrac{1}{3n} + \dfrac{1}{4n} = \dfrac{7}{24}$              $[2]$

28  $\dfrac{x+3}{4} = \dfrac{x-3}{5} + 2$                         $[13]$

29  $\dfrac{3t}{20} = \dfrac{6-t}{12} + \dfrac{2t}{15} - \dfrac{3}{2}$   $[-10]$

30  $\dfrac{y}{5} + \dfrac{7}{20} = \dfrac{5-y}{4}$              $[2]$

31  $\dfrac{v-2}{2v-3} = \dfrac{1}{3}$                            $[3]$

32  $\dfrac{2}{a-3} = \dfrac{3}{2a+1}$                            $[-11]$

33  $\dfrac{1}{3m-2} + \dfrac{1}{5m+3} = 0$                       $\left[-\dfrac{1}{8}\right]$

34  $\dfrac{x}{4} - \dfrac{x+6}{5} = \dfrac{x+3}{2}$             $[-6]$

35  $\dfrac{2c-3}{4} - \dfrac{1-c}{5} - 1 = \dfrac{2c+3}{3} + \dfrac{43}{60}$   $[110]$

36  $3\sqrt{t} = 9$                                               $[9]$

37  $2\sqrt{y} = 5$                                               $\left[6\dfrac{1}{4}\right]$

38  $4 = \sqrt{\left(\dfrac{3}{a}\right)} + 3$                    $[3]$

39  $\dfrac{3\sqrt{x}}{1-\sqrt{x}} = -6$                          $[4]$

40  $10 = 5\sqrt{\left(\dfrac{x}{2}-1\right)}$                    $[10]$

41  $16 = \dfrac{t^2}{9}$                                         $[\pm12]$

42 $\sqrt{\left(\dfrac{y+2}{y-2}\right)} = \dfrac{1}{2}$ $\qquad\qquad\qquad\left[-3\tfrac{1}{3}\right]$

43 $\dfrac{6}{a} = \dfrac{2a}{3}$ $\qquad$ [±3] $\qquad$ 44 $\dfrac{11}{2} = 5 + \dfrac{8}{x^2}$ $\qquad$ [±4]

*Practical problems involving simple equations*

45 A formula used for calculating resistance of a cable is $R = (\rho l)/a$.
   Given $R = 1.25$, $l = 2500$ and $a = 2 \times 10^{-4}$ find the value of $\rho$.

$[10^{-7}]$

46 Force $F$ newtons is given by $F = ma$, where $m$ is the mass in kilograms and $a$ is the acceleration in metres per second squared. Find the acceleration when a force of 4 kN is applied to a mass of 500 kg.

$[8 \text{ m/s}^2]$

47 $PV = mRT$ is the characteristic gas equation. Find the value of $m$ when $P = 100 \times 10^3$, $V = 3.00$, $R = 288$ and $T = 300$.

$[3.472]$

48 When three resistors $R_1$, $R_2$ and $R_3$ are connected in parallel the total resistance $R_T$ is determined from

$$\dfrac{1}{R_T} = \dfrac{1}{R_1} + \dfrac{1}{R_2} + \dfrac{1}{R_3}$$

   (a) Find the total resistance when $R_1 = 3\Omega$, $R_2 = 6\Omega$ and $R_3 = 18\Omega$.
   (b) Find the value of $R_3$ given that $R_T = 3\Omega$, $R_1 = 5\Omega$ and $R_2 = 10\Omega$

[(a) 1.8 $\Omega$; (b) 30 $\Omega$]

49 Five pens and two rulers cost 94p. If a ruler costs 5p more than a pen, find the cost of each.

[12p; 17p]

50 Ohm's law may be represented by $I = V/R$, where $I$ is the current in amperes, $V$ is the voltage in volts and $R$ is the resistance in ohms. A soldering iron takes a current of 0.30 A from a 240 V supply. Find the resistance of the element.

[800 $\Omega$]

51 A rectangle has a length of 20 cm and a width $b$ cm. When its width is reduced by 4 cm its area becomes 160 cm$^2$. Find the original width and area of the rectangle.

[12 cm; 240 cm$^2$]

52 Given $R_2 = R_1(1+\alpha t)$, find $\alpha$ given $R_1 = 5.0$, $R_2 = 6.03$ and $t = 51.5$. [0.004]

53 If $v^2 = u^2 + 2as$, find $u$ given $v = 24$, $a = -40$ and $s = 4.05$. [30]

54 The relationship between the temperature on a Fahrenheit scale and that on a Celsius scale is given by $F = \frac{9}{5}C + 32$. Express 113 °F in degrees Celsius.

[45 °C]

55 If $t = 2\pi\sqrt{(w/Sg)}$, find the value of $S$ given $w = 1.219$, $g = 9.81$ and $t = 0.3132$.

[50]

56 Two joiners and five mates earn £484 between them for a particular job. If a joiner earns £18 more than a mate, calculate the earnings for a joiner and for a mate.

[£82; £64]

57 An alloy contains 60% by weight of copper, the remainder being zinc. How much copper must be mixed with 50 kg of this alloy to give an alloy containing 75% copper?

[30 kg]

58 A rectangular laboratory has a length equal to one and a half times its width and a perimeter of 40 m. Find its length and width.

[12 m; 8 m]

76

# 6 Simultaneous equations

## A. MAIN POINTS CONCERNED WITH SIMULTANEOUS EQUATIONS

1   Only one equation is necessary when finding the value of a **single unknown quantity** (as with simple equations in chapter 5).
2   When an equation contains **two unknown quantities** it has an infinite number of solutions. When two equations are available connecting the same two unknown values then a unique solution is possible. Similarly, for three unknown quantities it is necessary to have three equations in order to solve for a particular value of each of the unknown quantities, and so on.
3   Equations which have to be solved together to find the unique values of the unknown quantities, which are true for each of the equations, are called **simultaneous equations**.
4   There are two methods of solving simultaneous equations analytically:
(a) by **substitution**, and (b) by **elimination**.

## B. WORKED PROBLEMS ON SIMULTANEOUS EQUATIONS IN TWO UNKNOWNS

*Problem 1* Solve the following equations for $x$ and $y$, (a) by substitution, and (b) by elimination:

$$x + 2y = -1 \tag{1}$$
$$4x - 3y = 18 \tag{2}$$

(a) *By substitution:*
From equation (1): $x = -1 - 2y$
Substituting this expression for $x$ into equation (2) gives:

$$4(-1 - 2y) - 3y = 18$$

This is now a simple equation in $y$.
Removing the bracket gives: $\quad -4 - 8y - 3y = 18$
$$-11y = 18 + 4 = 22$$
$$y = \frac{22}{-11} = -2$$

Substituting $y = -2$ into equation (1) gives: $\quad x + 2(-2) = -1$
$$x - 4 = -1$$
$$x = -1 + 4 = 3.$$

Thus $x = 3$ **and** $y = -2$ **is the solution to the simultaneous equations**
(Check: In equation (2), since $x = 3$ and $y = -2$, LHS $= 4(3) - 3(-2)$
$$= 12 + 6 = 18 = \text{RHS})$$

77

(b) *By elimination*

$$x + 2y = -1 \tag{1}$$
$$4x - 3y = 18 \tag{2}$$

If equation (1) is multiplied throughout by 4 the coefficient of $x$ will be the same as in equation (2), giving:

$$4x + 8y = -4 \tag{3}$$

Subtracting equation (3) from equation (2) gives:

$$4x - 3y = 18 \tag{2}$$
$$\underline{4x + 8y = -4} \tag{3}$$
$$0 \;\; -11y = 22$$

Hence $y = \dfrac{22}{-11} = -2$

(Note, in the above subtraction, $18 - -4 = 18 + 4 = 22$)
Substituting $y = -2$ into either equation (1) or equation (2) will give $x = 3$ as in method (a). The solution $x = 3, y = -2$ is the only pair of values that satisfies both of the original equations.

*Problem 2* Solve $\quad 7x - 2y = 26 \hfill (1)$
$$\qquad\qquad\qquad 6x + 5y = 29 \hfill (2)$$

When equation (1) is multiplied by 5 and equation (2) by 2 the coefficients of $y$ in each equation are numerically the same, i.e. 10, but are of opposite sign.

| | | |
|---|---|---|
| $5 \times$ equation (1) gives: | $35x - 10y = 130$ | (3) |
| $2 \times$ equation (2) gives: | $\underline{12x + 10y = \phantom{0}58}$ | (4) |
| Adding equations (3) and (4) gives: | $47x + \phantom{0}0 \phantom{0}= 188$ | |

Hence $x = \dfrac{188}{47} = 4$

[Note that when the signs of common coefficients are **different** the two equations are **added**, and when the signs of common coefficients are the same the two equations are **subtracted** (as in *Problem 1*).]

Substituting $x = 4$ into equation (1) gives: $\quad 7(4) - 2y = 26$
$$28 \;\; -2y = 26$$
$$28 \;\; -26 = 2y$$
$$2 = 2y$$
$$\text{Hence} \quad y = 1$$

Checking, by substituting $x = 4, y = 1$ into equation (2), gives:

LHS $= 6(4) + 5(1) = 24 + 5 = 29 =$ RHS

**Thus the solution is $x = 4, y = 1$**, since these values maintain the equality when substituted in both equations.

**Problem 3** Solve:   $3p = 2q$             (1)
                       $4p+q+11 = 0$      (2)

Rearranging gives:                    $3p-2q = 0$       (3)
                                   $4p+\ q = -11$    (4)

Multiplying equation (4) by 2 gives:     $8p+2q = -22$    (5)
Adding equations (3) and (5) gives:     $11p+0 = -22$

$$p \;=\; \frac{-22}{11} \;=\; -2$$

Substituting $p = -2$ into equation (1) gives:   $3(-2) = 2q$
                                            $-6 = 2q$

$$q \;=\; \frac{-6}{2} \;=\; -3$$

Checking, by substituting $p = -2$ and $q = -3$ into equation (2), gives:

LHS $= 4(-2)+(-3)+11 = -8-3+11 = 0 =$ RHS

**Hence the solution is $p = -2, q = -3$.**

**Problem 4** Solve $\dfrac{x}{8}+\dfrac{5}{2}= y$            (1)

$$13-\frac{y}{3} = 3x \qquad\qquad (2)$$

Whenever fractions are involved in simultaneous equations it is usual to firstly remove them. Thus, multiplying equation (1) by 8 gives:

$$8\left(\frac{x}{8}\right) + 8\left(\frac{5}{2}\right) = 8y$$

i.e. $x+20 = 8y$                                          (3)

Multiplying equation (2) by 3 gives:       $39-y = 9x$     (4)
Rearranging equations (3) and (4) gives:   $x-8y = -20$   (5)
                                        $9x+y = 39$    (6)
Multiplying equation (6) by 8 gives:       $72x+8y = 312$  (7)
Adding equations (5) and (7) gives:       $73x+\ 0 = 292$

$$x \;=\; \frac{292}{73} = 4$$

Substituting $x = 4$ into equation (5) gives:   $4-8y = -20$
                                         $4+20 = 8y$
                                            $24 = 8y$

$$y = \frac{24}{8} = 3$$

Checking, substituting $x = 4, y = 3$ in the original equations, gives:

*Equation (1)*:   LHS $= \dfrac{4}{8} + \dfrac{5}{2} = \dfrac{1}{2} + 2\dfrac{1}{2} = 3 = y =$ RHS

*Equation (2)*:    LHS = $13 - \dfrac{3}{3} = 13-1 = 12$

RHS = $3x = 3(4) = 12$

**Hence the solution is $x = 4, y = 3$**

---

*Problem 5*  Solve $2.5x+0.75-3y = 0$
$1.6x = 1.08-1.2y$

---

It is often easier to initially remove decimal fractions. Thus multiplying equations (1) and (2) by 100 gives:

|  |  |  |
|---|---|---|
|  | $250x+75-300y = 0$ | (1) |
|  | $160x = 108-120y$ | (2) |
| Rearranging gives: | $250x-300y = -75$ | (3) |
|  | $160x+120y = 108$ | (4) |
| Multiplying equation (3) by 2 gives: | $500x-600y = -150$ | (5) |
| Multiplying equation (4) by 5 gives: | $800x+600y = 540$ | (6) |
| Adding equations (5) and (6) gives: | $1\,300x+0 = 390$ |  |

$$x = \frac{390}{1\,300} = \frac{39}{130} = \frac{3}{10} = 0.3$$

Substituting $x = 0.3$ into equation (1) gives: $250(0.3)+75-300y = 0$

$$75+75 = 300y$$
$$150 = 300y$$
$$y = \frac{150}{300} = 0.5$$

Checking $x = 0.3, y = 0.5$ into equation (2) gives:

LHS = $160(0.3) = 48$     RHS = $108-120(0.5) = 108-60 = 48$

**Hence the solution is $x = 0.3, y = 0.5$**

---

*Problem 6*  Solve $\dfrac{2}{x} + \dfrac{3}{y} = 7$        (1)

$\dfrac{1}{x} - \dfrac{4}{y} = -2$        (2)

---

In this type of equation a substitution can initially be made.

Let $\dfrac{1}{x} = a$ and $\dfrac{1}{y} = b$

| Thus equation (1) becomes: | $2a+3b = 7$ | (3) |
|---|---|---|
| and equation (2) becomes: | $a-4b = -2$ | (4) |
| Multiplying equation (4) by 2 gives: | $2a-8b = -4$ | (5) |
| Subtracting equation (5) from equation (3) gives: | $0+11b = 11$ |  |
| i.e. | $b = 1$ |  |
| Substituting $b = 1$ into equation (3) gives: | $2a+3 = 7$ |  |
|  | $2a = 7-3 = 4$ |  |
| i.e. | $a = 2$ |  |

Checking, substituting $a = 2, b = 1$ in equation (4), gives:

LHS $= 2-4(1) = 2-4 = -2 =$ RHS
Hence $a = 2, b = 1$

However, since $\dfrac{1}{x} = a$ then $x = \dfrac{1}{a} = \dfrac{1}{2}$

and since $\dfrac{1}{y} = b$ then $y = \dfrac{1}{b} = \dfrac{1}{1} = 1$

**Hence the solution is** $x = \dfrac{1}{2}, y = 1$, which may be checked in the original equations.

*Problem 7* Solve $\dfrac{1}{2a} + \dfrac{3}{5b} = 4$  (1)

$\qquad\qquad \dfrac{4}{a} + \dfrac{1}{2b} = 10.5$  (2)

Let $\dfrac{1}{a} = x$ and $\dfrac{1}{b} = y$

Then $\dfrac{x}{2} + \dfrac{3}{5}y = 4$  (3)

$\qquad 4x + \dfrac{1}{2}y = 10.5$  (4)

To remove fractions, equation (3) is multiplied by 10 giving:

$10\left(\dfrac{x}{2}\right) + 10\left(\dfrac{3}{5}y\right) = 10(4)$

$\qquad\qquad\qquad$ i.e. $\;5x + 6y = 40$  (5)

Multiplying equation (4) by 2 gives: $\qquad 8x + y = 21$  (6)
Multiplying equation (6) by 6 gives: $\quad 48x + 6y = 126$  (7)
Subtracting equation (5) from equation (7) gives: $\;43x + 0 = 86$

$\qquad\qquad\qquad\qquad\qquad\qquad\qquad x = \dfrac{86}{43} = 2$

Substituting $x = 2$ into equation (3) gives: $\quad \dfrac{2}{2} + \dfrac{3}{5}y = 4$

$\qquad\qquad\qquad\qquad\qquad\qquad\qquad \dfrac{3}{5}y = 4-1 = 3$

$\qquad\qquad\qquad\qquad\qquad\qquad\qquad y = \dfrac{5}{3}(3) = 5$

Since $\dfrac{1}{a} = x$ then $a = \dfrac{1}{x} = \dfrac{1}{2}$

and since $\dfrac{1}{b} = y$ then $b = \dfrac{1}{y} = \dfrac{1}{5}$

**Hence the solution is** $a = \dfrac{1}{2}, b = \dfrac{1}{5}$, which may be checked in the original equations.

**Problem 8** Solve $\dfrac{x-1}{3} + \dfrac{y+2}{5} = \dfrac{2}{15}$        (1)

$\dfrac{1-x}{6} + \dfrac{5+y}{2} = \dfrac{5}{6}$        (2)

Before equations (1) and (2) can be simultaneously solved, the fractions need to be removed and the equations rearranged.

Multiplying equation (1) by 15 gives:     $15\left(\dfrac{x-1}{3}\right) + 15\left(\dfrac{y+2}{5}\right) = 15\left(\dfrac{2}{15}\right)$

i.e.     $5(x-1) + 3(y+2) = 2$

$5x - 5 + 3y + 6 = 2$

$5x + 3y = 2 + 5 - 6$

Hence     $5x + 3y = 1$        (3)

Multiplying equation (2) by 6 gives:     $6\left(\dfrac{1-x}{6}\right) + 6\left(\dfrac{5+y}{2}\right) = 6\left(\dfrac{5}{6}\right)$

i.e.     $(1-x) + 3(5+y) = 5$

$1 - x + 15 + 3y = 5$

$-x + 3y = 5 - 1 - 15$

Hence     $-x + 3y = -11$        (4)

Thus the initial problem containing fractions can be expressed as:

$$5x + 3y = 1 \qquad (3)$$
$$-x + 3y = -11 \qquad (4)$$

Subtracting equation (4) from equation (3) gives:     $6x + 0 = 12$

$$x = \frac{12}{6} = 2$$

Substituting $x = 2$ into equation (3) gives:     $5(2) + 3y = 1$

$10 + 3y = 1$

$3y = 1 - 10 = -9$

$$y = \frac{-9}{3} = -3$$

Checking, substituting $x = 2$, $y = -3$ in equation (4), gives:

LHS $= -2 + 3(-3) = -2 - 9 = -11 =$ RHS

**Hence the solution is $x = 2$, $y = -3$, which may be checked in the original equations.**

*Further problems on simultaneous equations may be found in section C, problems 1 to 27, page 86.*

*Problem 9* The law concerning friction $F$ and load $L$ for an experiment is of the form $F = aL+b$, where $a$ and $b$ are constants. When $F = 5.6$, $L = 8.0$ and when $F = 4.4$, $L = 2.0$. Find the values of $a$ and $b$ and the value of $F$ when $L = 6.5$.

Substituting $F = 5.6$, $L = 8.0$ into $F = aL+b$ gives: $\quad 5.6 = 8.0a+b \qquad (1)$
Substituting $F = 4.4$, $L = 2.0$ into $F = aL+b$ gives: $\quad 4.4 = 2.0a+b \qquad (2)$
Subtracting equation (2) from equation (1) gives: $\quad 1.2 = 6.0a$

$$a = \frac{1.2}{6.0} = \frac{1}{5}$$

Substituting $a = \frac{1}{5}$ into equation (1) gives: $\qquad 5.6 = 8.0\left(\frac{1}{5}\right)+b$

$$5.6 = 1.6+b$$
$$5.6-1.6 = b$$
i.e. $\qquad b = 4$

Checking, substituting $a = \frac{1}{5}$, $b = 4$ in equation (2), gives:

RHS $= 2.0\left(\frac{1}{5}\right)+4 = 0.4+4 = 4.4 =$ LHS

**Hence $a = \frac{1}{5}$ and $b = 4$**

When $L = 6.5$, $F = aL+b = \frac{1}{5}(6.5)+4 = 1.3+4$ i.e. $F = \mathbf{5.30}$

*Problem 10* The equation of a straight line, of slope $m$ and intercept on the $y$-axis $c$, is $y = mx+c$. If a straight line passes through the point where $x = 1$ and $y = -2$, and also through the point where $x = 3\frac{1}{2}$ and $y = 10\frac{1}{2}$, find the values of the slope and the $y$-axis intercept.

Substituting $x = 1$ and $y = -2$ into $y = mx+c$ gives: $\quad -2 = m+c \qquad (1)$

Substituting $x = 3\frac{1}{2}$ and $y = 10\frac{1}{2}$ into $y = mx+c$ gives: $10\frac{1}{2} = 3\frac{1}{2}m+c \qquad (2)$

Subtracting equation (1) from equation (2) gives: $\qquad 12\frac{1}{2} = 2\frac{1}{2}m$

$$m = \frac{12\frac{1}{2}}{2\frac{1}{2}} = 5$$

Substituting $m = 5$ into equation (1) gives: $\qquad -2 = 5+c$
$$c = -2-5 = -7$$

Checking, substituting $m = 5$, $c = -7$ in equation (2), gives:

RHS $= 3\frac{1}{2}(5)+(-7) = 17\frac{1}{2}-7 = 10\frac{1}{2} =$ LHS

**Hence the slope, $m = 5$ and the $y$-axis intercept, $c = -7$**

*Problem 11* When Kirchhoff's laws are applied to a particular electrical circuit the currents $I_1$ and $I_2$ are connected by the equations:

$$27 = 1.5I_1 + 8(I_1 - I_2) \qquad (1)$$
$$-26 = 2I_2 - 8(I_1 - I_2) \qquad (2)$$

Solve the equations to find the values of currents $I_1$ and $I_2$

Removing brackets from equation (1) gives: $\quad 27 = 1.5I_1 + 8I_1 - 8I_2$

Rearranging gives: $\qquad\qquad\qquad 9.5I_1 - 8I_2 = 27 \qquad (3)$

Removing brackets from equation (2) gives: $\quad -26 = 2I_2 - 8I_1 + 8I_2$

Rearranging gives: $\qquad\qquad\qquad -8I_1 + 10I_2 = -26 \qquad (4)$

Multiplying equation (3) by 5 gives: $\qquad 47.5I_1 - 40I_2 = 135 \qquad (5)$

Multiplying equation (4) by 4 gives: $\qquad -32I_1 + 40I_2 = -104 \qquad (6)$

Adding equations (5) and (6) gives: $\qquad 15.5I_1 + 0 \quad = 31$

$$I_1 = \frac{31}{15.5} = 2$$

Substituting $I_1 = 2$ into equation (3) gives: $\quad 9.5(2) - 8I_2 = 27$

$$19 - 8I_2 = 27$$
$$19 - 27 = 8I_2$$
$$-8 = 8I_2$$
$$I_2 = -1$$

**Hence the solution is $I_1 = 2$ and $I_2 = -1$**, (which may be checked in the original equations)

*Problem 12* The distance $s$ metres from a fixed point of a vehicle travelling in a straight line with constant acceleration, $a$ m/s$^2$, is given by $s = ut + \frac{1}{2}at^2$, where $u$ is the initial velocity in m/s and $t$ the time in seconds. Determine the initial velocity and the acceleration given that $s = 42$ m when $t = 2$ s and $s = 144$ m when $t = 4$ s. Find also the distance travelled after 3 s

Substituting $s = 42$, $t = 2$ into $s = ut + \frac{1}{2}at^2$ gives: $\quad 42 = 2u + \frac{1}{2}a(2)^2$

i.e. $\qquad 42 = 2u + 2a \qquad (1)$

Substituting $s = 144$, $t = 4$ into $s = ut + \frac{1}{2}at^2$ gives: $\quad 144 = 4u + \frac{1}{2}a(4)^2$

i.e. $\qquad 144 = 4u + 8a \qquad (2)$

Multiplying equation (1) by 2 gives: $\qquad\qquad\qquad 84 = 4u + 4a \qquad (3)$

Subtracting equation (3) from equation (2) gives: $\qquad 60 = 0 + 4a$

$$a = \frac{60}{4} = 15$$

Substituting $a = 15$ into equation (1) gives: $\qquad 42 = 2u + 2(15)$

$$42 - 30 = 2u$$
$$u = \frac{12}{2} = 6$$

Substituting $a = 15, u = 6$ in equation (2) gives:

RHS $= 4(6)+8(15) = 24+120 = 144 =$ LHS

**Hence the initial velocity, $u = 6$ m/s and the acceleration, $a = 15$ m/s$^2$**

Distance travelled after 3 s is given by $s = ut+\frac{1}{2}at^2$, where $t = 3, u = 6$ and $a = 15$.

Hence $\quad s = 6(3)+\frac{1}{2}(15)(3)^2$

$\qquad = 18+67\frac{1}{2}$

i.e. **distance travelled after 3 s $= 85\frac{1}{2}$ m**

---

*Problem 13*  A craftsman and 4 labourers together earn £357 basic per week, whilst 4 craftsmen and 9 labourers together earn £952 basic per week. Determine the basic weekly wage of a craftsman and a labourer.

---

Let $C$ represent the wage of a craftsman and $L$ that of a labourer.

Thus $\quad C+4L = 357 \qquad\qquad\qquad\qquad\qquad\qquad\qquad (1)$

$\qquad\quad 4C+9L = 952 \qquad\qquad\qquad\qquad\qquad\qquad\qquad (2)$

Multiplying equation (1) by 4 gives: $\qquad\qquad 4C+16L = 1\ 428 \qquad (3)$

Subtracting equation (2) from equation (3) gives: $\qquad 7L = 476$

$$L = \frac{476}{7} = 68$$

Substituting $L = 68$ into equation (1) gives: $\qquad C+4(68) = 357$

$\qquad\qquad\qquad\qquad\qquad\qquad\qquad\qquad\qquad\quad C+272 = 357$

$\qquad\qquad\qquad\qquad\qquad\qquad\qquad\qquad\qquad\quad\ \ C = 357-272 = 85$

Checking, substituting $C = 85, L = 68$ in equation (2), gives:

LHS $= 4(85)+9(68) = 340+612 = 952 =$ RHS

**Thus the solution is that the basic weekly wage of a craftsman is £85 and that of a labourer is £68.**

---

*Problem 14*  The resistance $R\ \Omega$ of a length of wire at $t°C$ is given by $R = R_0(1+\alpha t)$, where $R_0$ is the resistance at $0°C$ and $\alpha$ is the temperature coefficient of resistance in $/°C$. Find the values of $\alpha$ and $R_0$ if $R = 30$ ohms at $50°C$ and $R = 35\ \Omega$ at $100°C$

---

Substituting $R = 30, t = 50$ into $R = R_0(1+\alpha t)$ gives: $\quad 30 = R_0(1+50\alpha) \qquad (1)$

Substituting $R = 35, t = 100$ into $R = R_0(1+\alpha t)$ gives: $\quad 35 = R_0(1+100\alpha) \qquad (2)$

Although these equations may be solved by the conventional substitution method, an easier way is to eliminate $R_0$ by division. Thus, dividing equation (1) by equation (2) gives:

$$\frac{30}{35} = \frac{R_0(1+50\alpha)}{R_0(1+100\alpha)} = \frac{1+50\alpha}{1+100\alpha}$$

'Cross-multiplying' gives:
$$30(1+100\alpha) = 35(1+50\alpha)$$
$$30+3\,000\alpha = 35+1\,750\alpha$$
$$3\,000\alpha-1\,750\alpha = 35-30$$
$$1\,250\alpha = 5$$

i.e. $\qquad \alpha = \dfrac{5}{1\,250} = \dfrac{1}{250}$ or $0.004$

Substituting $\alpha = \dfrac{1}{250}$ into equation (1) gives: $30 = R_0\left[1 + \dfrac{1}{250}(50)\right]$

$$30 = R_0(1.2)$$

$$R_0 = \frac{30}{1.2} = 25$$

Checking, substituting $\alpha = \dfrac{1}{250}$, $R_0 = 25$ in equation (2), gives:

RHS $= 25\left[1 + \dfrac{1}{250}(100)\right] = 25(1.4) = 35 =$ LHS

**Thus the solution is $\alpha = 0.004/°$C and $R_0 = 25\ \Omega$.**

*Further examples on practical problems involving simultaneous equations in two unknowns may be found in section C following, problems 28 to 35, page 88.*

---

## C. FURTHER PROBLEMS ON SIMULTANEOUS EQUATIONS IN TWO UNKNOWNS

In *Problems 1 to 27*, solve the simultaneous equations and verify the results.

1  $a+b = 7$
   $a-b = 3$
   $\hfill [a = 5; b = 2]$

2  $2x+5y = 7$
   $x+3y = 4$
   $\hfill [x = 1; y = 1]$

3  $3s+2t = 12$
   $4s-t = 5$
   $\hfill [s = 2; t = 3]$

4  $3x-2y = 13$
   $2x+5y = -4$
   $\hfill [x = 3; y = -2]$

5  $5m-3n = 11$
   $3m+n = 8$
   $\hfill \left[m = 2\tfrac{1}{2}; n = \tfrac{1}{2}\right]$

6  $8a-3b = 51$
   $3a+4b = 14$
   $\hfill [a = 6; b = -1]$

7  $5x = 2y$
   $3x+7y = 41$
   $\hfill [x = 2; y = 5]$

8   $5c = 1-3d$
     $2d+c+4 = 0$

$$[c = 2; d = -3]$$

9   $7p+11+2q = 0$
     $-1 = 3q-5p$

$$[p = -1; q = -2]$$

10  $\dfrac{x}{2}+\dfrac{y}{3} = 4; \quad \dfrac{x}{6}-\dfrac{y}{9} = 0$

$$[x = 4; y = 6]$$

11  $\dfrac{a}{2} - 7 = -2b; \quad 12 = 5a + \dfrac{2}{3}b$

$$[a = 2; b = 3]$$

12  $\dfrac{3}{2}s-2t = 8; \quad \dfrac{s}{4} + 3t = -2$

$$[s = 4; t = -1]$$

13  $\dfrac{x}{5}+\dfrac{2y}{3} = \dfrac{49}{15}; \quad \dfrac{3x}{7}-\dfrac{y}{2}+\dfrac{5}{7} = 0$

$$[x = 3; y = 4]$$

14  $v-1 = \dfrac{u}{12}; \quad u+\dfrac{v}{4}-\dfrac{25}{2} = 0$

$$[u = 12; v = 2]$$

15  $1.5x-2.2y = -18$
     $2.4x+0.6y = 33$

$$[x = 10; y = 15]$$

16  $3b-2.5a = 0.45$
     $1.6a+0.8b = 0.8$

$$[a = 0.30; b = 0.40]$$

17  $10.1+1.7y = 0.8x$
     $2.5x+1.5+1.3y = 0$

$$[x = 2; y = -5]$$

18  $0.4b-0.7 = 0.5a$
     $1.2a-3.6 = 0.3b$

$$[a = 5; b = 8]$$

19  $2.30c-1.70d = 9.11$
     $3.68+8.80c+4.20d = 0$

$$[c = 1.3; d = -3.6]$$

20  $\dfrac{3}{x}+\dfrac{2}{y} = 14; \quad \dfrac{5}{x}-\dfrac{3}{y} = -2$

$$\left[x = \dfrac{1}{2}; y = \dfrac{1}{4}\right]$$

21  $\dfrac{4}{a}-\dfrac{3}{b} = 18; \quad \dfrac{2}{a}+\dfrac{5}{b} = -4$

$$\left[a = \dfrac{1}{3}; b = -\dfrac{1}{2}\right]$$

22  $\dfrac{1}{2p}+\dfrac{3}{5q} = 5; \quad \dfrac{5}{p} - \dfrac{1}{2q} = \dfrac{35}{2}$

$$\left[p = \dfrac{1}{4}; q = \dfrac{1}{5}\right]$$

23  $\dfrac{5}{x}+\dfrac{3}{y} = 1.1; \quad \dfrac{3}{x} - \dfrac{7}{y} = -1.1$

$$[x = 10; y = 5]$$

24  $\dfrac{c+1}{4} - \dfrac{d+2}{3} +1 = 0; \quad \dfrac{1-c}{5} + \dfrac{3-d}{4} + \dfrac{13}{20} = 0$

$$[c = 3; d = 4]$$

25  $\dfrac{3r+2}{5} - \dfrac{2s-1}{4} = \dfrac{11}{5}; \quad \dfrac{3+2r}{4} + \dfrac{5-s}{3} = \dfrac{15}{4}$

$$\left[r = 3; s = \dfrac{1}{2}\right]$$

26  $\dfrac{5}{x+y} = \dfrac{20}{27}; \quad \dfrac{4}{2x-y} = \dfrac{16}{33}$

$$\left[x = 5; y = 1\dfrac{3}{4}\right]$$

27  If $5x-\dfrac{3}{y} = 1$ and $x+\dfrac{4}{y} = \dfrac{5}{2}$ find the value of $\dfrac{xy+1}{y}$

$$[1]$$

*Practical problems involving simultaneous equations in two unknowns*

28  In a system of pulleys, the effort $P$ required to raise a load $W$ is given by $P = aW + b$, where $a$ and $b$ are constants. If $W = 40$ when $P = 12$ and $W = 90$ when $P = 22$, find the values of $a$ and $b$.

$$\left[ a = \frac{1}{5}; b = 4 \right]$$

29  Applying Kirchhoff's laws to an electrical circuit produces the following equations:

$$5 = 0.2I_1 + 2(I_1 - I_2)$$
$$12 = 3I_2 + 0.4I_2 - 2(I_1 - I_2).$$

Determine the values of currents $I_1$ and $I_2$.

$$[I_1 = 6.47; I_2 = 4.62]$$

30  Velocity $v$ is given by the formula $v = u + at$. If $v = 20$ when $t = 2$ and $v = 40$ when $t = 7$ find the values of $u$ and $a$. Hence find the velocity when $t = 3.5$.

$$[u = 12; a = 4; v = 26]$$

31  Three new cars and 4 new vans supplied to a dealer together cost £19 600 and 5 new cars and 2 new vans of the same models cost £21 000. Find the cost of a car and a van.

$$[£3\ 200; £2\ 500]$$

32  $y = mx + c$ is the equation of a straight line of slope $m$ and $y$-axis intercept $c$. If the line passes through the point where $x = 2$ and $y = 2$, and also through the point where $x = 5$ and $y = \frac{1}{2}$, find the slope and $y$-axis intercept of the straight line.

$$\left[ m = -\frac{1}{2}; c = 3 \right]$$

33  The resistance $R$ ohms of copper wire at $t°C$ is given by $R = R_0(1 + \alpha t)$, where $R_0$ is the resistance at $0°C$ and $\alpha$ is the temperature coefficient of resistance. If $R = 25.44\ \Omega$ at $30°C$ and $R = 32.17\ \Omega$ at $100°C$, find $\alpha$ and $R_0$.

$$[\alpha = 0.004\ 26; R_0 = 22.56\ \Omega]$$

34  The molar heat capacity of a solid compound is given by the equation $c = a + bT$. When $c = 52$, $T = 100$ and when $c = 172$, $T = 400$. Find the values of $a$ and $b$.

$$[a = 12; b = 0.40]$$

35  In an engineering process two variables $p$ and $q$ are related by: $q = ap + b/p$, where $a$ and $b$ are constants. Evaluate $a$ and $b$ if $q = 13$ when $p = 2$ and $q = 22$ when $p = 5$.

$$[a = 4; b = 10]$$

# 7 Evaluation and transposition of formulae

## A. MAIN POINTS CONCERNED WITH EVALUATION AND TRANSPOSITION OF FORMULAE

1   The statement $v = u + at$ is said to be a **formula** for $v$ in terms of $u$, $a$ and $t$. $v$, $u$, $a$ and $t$ are called **symbols**. The single term on the left hand side of the equation, $v$, is called the **subject of the formula**.

2   Provided values are given for all the symbols in a formula except one, the remaining symbol can be made the subject of the formula and may be evaluated by using tables, slide rules or calculators.

3   When a symbol other than the subject is required to be calculated it is usual to rearrange the formula to make a new subject. This rearranging process is called **transposing the formula** or **transposition**.

4   The rules used for transposition of formulae are the same as those used for the solution of simple equations (see chapter 5)—basically, **that the equality of an equation must be maintained.**

## B. WORKED PROBLEMS ON EVALUATION AND TRANSPOSITION OF FORMULAE

(a) EVALUATION OF FORMULAE

(The actual method of calculation in problems 1–9 is not specified; tables, slide-rules or calculators may be used.)

*Problem 1*   In an electrical circuit the voltage $V$ is given by Ohm's law, i.e. $V = IR$. Find, correct to 4 significant figures, the voltage when $I = 5.36$ A and $R = 14.76\ \Omega$

$V = IR = (5.36)(14.76)$

Hence **voltage $V = 79.11$ V**, correct to 4 significant figures

*Problem 2*   The surface area $A$ of a hollow cone is given by $A = \pi rl$. Determine the surface area when $r = 3.0$ cm, $l = 8.5$ cm and $\pi = 3.14$

$A = \pi rl = (3.14)(3.0)(8.5)\ \text{cm}^2$

Hence surface area $A = 80.07\ \text{cm}^2$

*Problem 3* Velocity $v$ is given by $v = u+at$. If $u = 9.86$ m/s, $a = 4.25$ m/s$^2$ and $t = 6.84$ s, find $v$, correct to 3 significant figures

$$v = u+at = 9.86+(4.25)(6.84)$$
$$= 9.86+29.07$$
$$= 38.93$$

**Hence velocity $v = 38.9$ m/s, correct to 3 significant figures**

*Problem 4* The area, $A$, of a circle is given by $A = \pi r^2$. Determine the area correct to 2 decimal places, given $\pi = 3.142$ and $r = 5.23$ m

$$A = \pi r^2 = (3.142)(5.23)^2$$
$$= (3.142)(27.35)$$

**Hence area, $A = 85.94$ m$^2$, correct to 2 decimal places**

*Problem 5* The power $P$ watts dissipated in an electrical circuit may be expressed by the formula $P = V^2/R$. Evaluate the power, correct to 3 significant figures, given that $V = 17.48$ V and $R = 36.12$ $\Omega$.

$$P = \frac{V^2}{R} = \frac{(17.48)^2}{36.12} = \frac{305.6}{36.12}$$

**Hence power, $P = 8.46$ W, correct to 3 significant figures**

*Problem 6* The volume $V$ cm$^3$ of a right circular cone is given by $V = \frac{1}{3}\pi r^2 h$. Given that $r = 4.321$ cm, $h = 18.35$ cm and $\pi = 3.142$, find the volume correct to 4 significant figures

$$V = \frac{1}{3}\pi r^2 h = \frac{1}{3}(3.142)(4.321)^2(18.35)$$
$$= \frac{1}{3}(3.142)(18.67)(18.35)$$

**Hence volume, $V = 358.8$ cm$^3$, correct to 4 significant figures**

*Problem 7* Force $F$ newtons is given by the formula $F = (Gm_1m_2)/d^2$, where $m_1$ and $m_2$ are masses, $d$ their distance apart and $G$ is a constant. Find the value of the force given that $G = 6.67 \times 10^{-11}$, $m_1 = 7.36$, $m_2 = 15.5$ and $d = 22.6$. Express the answer in standard form, correct to 3 significant figures.

$$F = \frac{Gm_1m_2}{d^2} = \frac{(6.67 \times 10^{-11})(7.36)(15.5)}{(22.6)^2} = \frac{(6.67)(7.36)(15.5)}{(10^{11})(510.8)} = \frac{1.490}{10^{11}}$$

**Hence force $F = 1.49 \times 10^{-11}$ newtons, correct to 3 significant figures**

*Problem 8* The time of swing, $t$ seconds, of a simple pendulum is given by $t = 2\pi\sqrt{(l/g)}$. Determine the time, correct to 3 decimal places, given that $\pi = 3.142, l = 12.0$ and $g = 9.81$

$$t = 2\pi\sqrt{\left(\frac{l}{g}\right)} = (2)(3.142) \quad \sqrt{\left(\frac{12.0}{9.81}\right)}$$
$$= (2)(3.142)\sqrt{(1.223)}$$
$$= (2)(\quad 142)(1.106)$$

**Hence time $t = 6.950$ seconds, correct to 3 decimal places**

*Problem 9* Resistance, $R\ \Omega$, varies with temperature according to the formula $R = R_0(1+\alpha t)$. Evaluate $R$, correct to 3 significant figures, given $R_0 = 14.59$, $\alpha = 0.004\ 3$ and $t = 80$

$$R = R_0(1+\alpha t) = 14.59[1+(0.004\ 3)(80)]$$
$$= 14.59(1+0.344)$$
$$= 14.59(1.344)$$

**Hence resistance, $R = 19.6\ \Omega$, correct to 3 significant figures**

*Further problems on evaluating formulae may be found in section C, problems 1 to 12, page 98.*

(b) TRANSPOSITION OF FORMULAE

*Problem 10* Transpose $p = q+r+s$ to make $r$ the subject

The aim is to obtain $r$ on its own on the left hand side (LHS) of the equation. Changing the equation around so that $r$ is on the LHS gives:

$$q+r+s = p \tag{1}$$

Subtracting $(q+s)$ from both sides of the equation gives:

$$q+r+s-(q+s) = p-(q+s)$$

Thus $\quad q+r+s-q-s = p-q-s$
i.e. $\qquad r = p-q-s \tag{2}$

It is shown with simple equations (chapter 5), that a quantity can be moved from one side of an equation to the other with an appropriate change of sign. Thus equation (2) follows immediately from equation (1) above.

*Problem 11* If $a+b = w-x+y$, express $x$ as the subject

Rearranging gives: $\quad w-x+y = a+b$ and $\quad -x = a+b-w-y$

Multiplying both sides by $-1$ gives:

$$(-1)(-x) = (-1)(a+b-w-y) \quad \text{i.e.} \quad x = -a-b+w+y$$

The result of multiplying each side of the equation by $-1$ is to change all the signs in the equation.

It is conventional to express answers with positive quantities first. Hence rather than $x = -a-b+w+y$, $x = w+y-a-b$, since the order of terms connected by $+$ and $-$ signs is immaterial.

*Problem 12* Transpose $v = f\,\lambda$ to make $\lambda$ the subject.

Rearranging gives: $\qquad f\lambda = v$

Dividing both sides by $f$ gives: $\quad \dfrac{f\lambda}{f} = \dfrac{v}{f} \quad$ i.e. $\quad \lambda = \dfrac{v}{f}$

*Problem 13* When a body falls freely through a height $h$, the velocity $v$ is given by $v^2 = 2gh$. Express this formula with $h$ as the subject.

Rearranging gives: $\qquad 2gh = v^2$

Dividing both sides by $2g$ gives: $\quad \dfrac{2gh}{2g} = \dfrac{v^2}{2g} \quad$ i.e. $\quad h = \dfrac{v^2}{2g}$

*Problem 14* If $A = B/C$, rearrange to make $B$ the subject.

Rearranging gives: $\qquad \dfrac{B}{C} = A$

Multiplying both sides by $C$ gives: $\quad C\left(\dfrac{B}{C}\right) = C(A)$

$$\text{Hence} \quad \mathbf{B} = CA$$

*Problem 15* Transpose $a = \dfrac{F}{m}$ for $m$

Rearranging gives: $\qquad \dfrac{F}{m} = a$

Multiplying both sides by $m$ gives: $\quad m\left(\dfrac{F}{m}\right) = m(a)$ i.e. $F = ma$

Rearranging gives: $\qquad ma = F$

Dividing both sides by $a$ gives: $\quad \dfrac{ma}{a} = \dfrac{F}{a} \quad$ i.e. $\quad m = \dfrac{F}{a}$

(i)  Rearranging gives: $\qquad\qquad\qquad\qquad \dfrac{\rho l}{a} = R$

Multiplying both sides by $a$ gives: $\quad a\left(\dfrac{\rho l}{a}\right) = a(R)$   i.e. $\rho l = aR$

Rearranging gives: $\qquad\qquad\qquad\quad aR = \rho l$

Dividing both sides by $R$ gives: $\qquad \dfrac{aR}{R} = \dfrac{\rho l}{R}$   i.e. $a = \dfrac{\rho l}{R}$

(ii) $\rho l/a = R$

Multiplying both sides by $a$ gives: $\quad \rho l \; = aR$

Dividing both sides by $\rho$ gives: $\qquad \dfrac{\rho l}{\rho} = \dfrac{aR}{\rho}$   i.e. $l \; = \dfrac{aR}{\rho}$

Rearranging gives: $\qquad\qquad\qquad u+\dfrac{ft}{m} = v$

and $\qquad\qquad\qquad\qquad\qquad \dfrac{ft}{m} = v-u$

Multiplying each side by $m$ gives: $\quad m\left(\dfrac{ft}{m}\right) = m(v-u)$

i.e. $\qquad\qquad ft = m(v-u)$

Dividing both sides by $t$ gives: $\qquad \dfrac{ft}{t} = \dfrac{m}{t}(v-u)$

i.e. $\quad f = \dfrac{m}{t}(v-u)$

Rearranging gives: $\qquad\qquad\qquad l_1(1+\alpha\theta) = l_2$

Removing the bracket gives: $\qquad\quad l_1+l_1\alpha\theta = l_2$

Rearranging gives: $\qquad\qquad\qquad\quad l_1\alpha\theta = l_2-l_1$

Dividing both sides by $l_1\theta$ gives: $\qquad \dfrac{l_1\alpha\theta}{l_1\theta} = \dfrac{l_2-l_1}{l_1\theta}$

i.e. $\qquad\qquad\qquad \alpha \; = \dfrac{l_2-l_1}{l_1\theta}$

**Problem 19** A formula for the distance moved by a body is given by $s = \frac{1}{2}(v+u)t$. Rearrange the formula to make $u$ the subject

Rearranging gives:

$$\frac{1}{2}(v+u)t = s$$

Multiplying both sides by 2 gives:

$$(v+u)t = 2s$$

Dividing both sides by $t$ gives:

$$\frac{(v+u)t}{t} = \frac{2s}{t}$$

i.e.

$$v+u = \frac{2s}{t}$$

Hence $u = \dfrac{2s}{t} - v$ or $\dfrac{2s-vt}{t}$

---

**Problem 20** A formula for kinetic energy is $k = \frac{1}{2}mv^2$. Transpose the formula to make $v$ the subject

Rearranging gives: $\frac{1}{2}mv^2 = k$

Whenever the prospective new subject is a squared term, that term is isolated on the LHS, and then the square root of both sides of the equation is taken.

Multiplying both sides by 2 gives: $mv^2 = 2k$

Dividing both sides by $m$ gives:

$$\frac{mv^2}{m} = \frac{2k}{m}$$

i.e.

$$v^2 = \frac{2k}{m}$$

Taking the square root of both sides gives: $\sqrt{v^2} = \sqrt{\left(\dfrac{2k}{m}\right)}$

i.e.

$$v = \sqrt{\left(\dfrac{2k}{m}\right)}$$

---

**Problem 21** In a right angled triangle having sides $x$, $y$ and hypotenuse $z$, Pythagoras' theorem states $z^2 = x^2+y^2$. Transpose the formula to find $x$

Rearranging gives:

$$x^2+y^2 = z^2$$

and

$$x^2 = z^2-y^2$$

Taking the square root of both sides gives: $x = \sqrt{(z^2-y^2)}$

Whenever the prospective new subject is within a square root sign, it is best to isolate that term on the LHS and then to square both sides of the equation.

Rearranging gives:
$$2\pi\sqrt{\left(\frac{l}{g}\right)} = t$$

Dividing both sides by $2\pi$ gives:
$$\sqrt{\left(\frac{l}{g}\right)} = \frac{t}{2\pi}$$

Squaring both sides gives:
$$\frac{l}{g} = \left(\frac{t}{2\pi}\right)^2 = \frac{t^2}{4\pi^2}$$

Cross-multiplying, i.e. multiplying each term by $4\pi^2 g$, gives:

$$4\pi^2 l = gt^2$$
$$\text{or} \quad gt^2 = 4\pi^2 l$$

Dividing both sides by $t^2$ gives:
$$\frac{gt^2}{t^2} = \frac{4\pi^2 l}{t^2}$$

i.e.
$$g = \frac{4\pi^2 l}{t^2}$$

*Problem 23* The impedance of an ac circuit is given by $Z = \sqrt{(R^2 + X^2)}$. Make the reactance, $X$, the subject.

Rearranging gives:       $\sqrt{(R^2 + X^2)} = Z$
Squaring both sides gives:    $R^2 + X^2 = Z^2$
Rearranging gives:        $X^2 = Z^2 - R^2$
Taking the square root of both sides gives:   $X = \sqrt{(Z^2 - R^2)}$

*Problem 24* The volume $V$ of a hemisphere is given by $V = \frac{2}{3}\pi r^3$. Find $r$ in terms of $V$

Rearranging gives:
$$\frac{2}{3}\pi r^3 = V$$

Multiplying both sides by 3 gives:    $2\pi r^3 = 3V$

Dividing both sides by $2\pi$ gives:
$$\frac{2\pi r^3}{2\pi} = \frac{3V}{2\pi}$$

i.e.
$$r^3 = \frac{3V}{2\pi}$$

Taking the cube root of both sides gives:  $\sqrt[3]{r^3} = \sqrt[3]{\left(\frac{3V}{2\pi}\right)}$

i.e.
$$r = \sqrt[3]{\left(\frac{3V}{2\pi}\right)}$$

**Problem 25** Transpose the formula $p = \dfrac{a^2x + a^2y}{r}$, to make $a$ the subject.

Rearranging gives:

$$\frac{a^2x + a^2y}{r} = p$$

Multiplying both sides by $r$ gives:    $a^2x + a^2y = rp$

Factorising the LHS gives:    $a^2(x+y) = rp$

Dividing both sides by $(x+y)$ gives:    $\dfrac{a^2(x+y)}{(x+y)} = \dfrac{rp}{(x+y)}$

i.e.    $a^2 = \dfrac{rp}{(x+y)}$

Taking the square root of both sides gives:    $a = \sqrt{\left(\dfrac{rp}{x+y}\right)}$

**Problem 26** Make $b$ the subject of the formula $a = \dfrac{x-y}{\sqrt{(bd+be)}}$

Rearranging gives:

$$\frac{x-y}{\sqrt{(bd+be)}} = a$$

Multiplying both sides by $\sqrt{(bd+be)}$ gives:    $x-y = a\sqrt{(bd+be)}$

or    $a\sqrt{(bd+be)} = x-y$

Dividing both sides by $a$ gives:

$$\sqrt{(bd+be)} = \frac{x-y}{a}$$

Squaring both sides gives:

$$bd + be = \left(\frac{x-y}{a}\right)^2$$

Factorising the LHS gives:

$$b(d+e) = \left(\frac{x-y}{a}\right)^2$$

Dividing both sides by $(d+e)$ gives:

$$b = \frac{\left(\dfrac{x-y}{a}\right)^2}{(d+e)}$$

i.e.

$$b = \frac{(x-y)^2}{a^2(d+e)}$$

**Problem 27** If $cd = 3d + e - ad$, express $d$ in terms of $a$, $c$ and $e$

Rearranging to obtain the terms in $d$ on the LHS gives:    $cd - 3d + ad = e$

Factorising the LHS gives:    $d(c - 3 + a) = e$

Dividing both sides by $(c-3+a)$ gives:    $d = \dfrac{e}{c-3+a}$

Rearranging gives:
$$\frac{b}{1+b} = a$$

Multiplying both sides by $(1+b)$ gives: $\quad b = a(1+b)$

Removing the bracket gives: $\quad b = a+ab$

Rearranging to obtain terms in $b$ on the LHS gives:
$$b-ab = a$$

Factorising the LHS gives: $\quad b(1-a) = a$

Dividing both sides by $(1-a)$ gives:
$$b = \frac{a}{1-a}$$

*Problem 29* Transpose the formula $V = \dfrac{Er}{R+r}$ to make $r$ the subject

Rearranging gives:
$$\frac{Er}{R+r} = V$$

Multiplying both sides by $(R+r)$ gives: $\quad Er = V(R+r)$

Removing the bracket gives: $\quad Er = VR+Vr$

Rearranging to obtain terms in $r$ on the LHS gives: $\quad Er-Vr = VR$

Factorising gives: $\quad r(E-V) = VR$

Dividing both sides by $(E-V)$ gives:
$$r = \frac{VR}{E-V}$$

*Problem 30* Transpose the formula $y = pq^2/(r+q^2) -t$, to make $q$ the subject

Rearranging gives:
$$\frac{pq^2}{r+q^2} -t = y$$

and
$$\frac{pq^2}{r+q^2} = y+t$$

Multiplying both sides by $(r+q^2)$ gives: $\quad pq^2 = (r+q^2)(y+t)$

Removing brackets gives: $\quad pq^2 = ry+rt+q^2y+q^2t$

Rearranging to obtain terms in $q$ on the LHS gives:
$$pq^2-q^2y-q^2t = ry+rt$$

Factorising gives: $\quad q^2(p-y-t) = r(y+t)$

Dividing both sides by $(p-y-t)$ gives:
$$q^2 = \frac{r(y+t)}{(p-y-t)}$$

Taking the square root of both sides gives:
$$q = \sqrt{\left(\frac{r(y+t)}{(p-y-t)}\right)}$$

*Problem 31* Given that $\dfrac{D}{d} = \sqrt{\left(\dfrac{f+p}{f-p}\right)}$ , express $p$ in terms of $D$, $d$ and $f$

Rearranging gives:

$$\sqrt{\left(\frac{f+p}{f-p}\right)} = \frac{D}{d}$$

Squaring both sides gives:

$$\frac{f+p}{f-p} = \frac{D^2}{d^2}$$

Cross-multiplying, i.e. multiplying each term by $d^2(f-p)$, gives:

$$d^2(f+p) = D^2(f-p)$$

Removing brackets gives:

$$d^2 f + d^2 p = D^2 f - D^2 p$$

Rearranging, to obtain terms in $p$ on the LHS, gives:

$$d^2 p + D^2 p = D^2 f - d^2 f$$

Factorising gives:

$$p(d^2 + D^2) = f(D^2 - d^2)$$

Dividing both sides by $(d^2 + D^2)$ gives:

$$p = \frac{f(D^2 - d^2)}{(d^2 + D^2)}$$

*Further problems on transposing formulae may be found in section C following, problems 13–45, page 99.*

## C. FURTHER PROBLEMS ON EVALUATION AND TRANSPOSITION OF FORMULAE

*Evaluation of formulae*

1   The area $A$ of a rectangle is given by the formula $A = lb$. Evaluate the area when $l = 12.4$ cm and $b = 5.37$ cm.

$$[A = 66.59 \text{ cm}^2]$$

2   The circumference $C$ of a circle is given by the formula $C = 2\pi r$. Determine the circumference given $\pi = 3.14$ and $r = 8.40$ mm.

$$[C = 52.75 \text{ mm}]$$

3   A formula used in connection with gases is $R = (PV)/T$. Evaluate $R$ when $P = 1\ 500$, $V = 5$ and $T = 200$.

$$[R = 37.5]$$

4   The potential difference, $V$ volts, available at battery terminals is given by $V = E - Ir$. Evaluate $V$ when $E = 5.62$, $I = 0.70$ and $R = 4.30$.

$$[V = 2.61 \text{ V}]$$

5   Given force $F = \frac{1}{2}m(v^2 - u^2)$, find $F$ when $m = 18.3$, $v = 12.7$ and $u = 8.24$

$$[F = 854.5]$$

6   The current $I$ amperes flowing in a number of cells is given by $I = (nE)/(R+nr)$. Evaluate the current when $n = 36$, $E = 2.20$, $R = 2.80$ and $r = 0.50$

$$[I = 3.81 \text{ A}]$$

7   The time, $t$ seconds, of oscillation for a simple pendulum is given by $t = 2\pi\sqrt{(l/g)}$. Determine the time when $\pi = 3.142$, $l = 54.32$ and $g = 9.81$

$$[t = 14.79 \text{ s}]$$

8   Energy, $E$ joules, is given by the formula $E = \frac{1}{2}LI^2$. Evaluate the energy when $L = 5.5$ and $I = 1.2$

$$[E = 3.96 \text{ J}]$$

9   The current $I$ amperes in an a.c. circuit is given by $I = \dfrac{V}{\sqrt{(R^2 + X^2)}}$
Evaluate the current when $V = 250$, $R = 11.0$ and $X = 16.2$

$$[I = 12.77 \text{ A}]$$

10 Distance $s$ metres is given by the formula $s = ut + \frac{1}{2}at^2$. If $u = 9.50$, $t = 4.60$ and $a = -2.50$, evaluate the distance

$$[s = 17.25 \text{ m}]$$

11 The area, $A$, of any triangle is given by $A = \sqrt{[s(s-a)(s-b)(s-c)]}$, where $s = \frac{a+b+c}{2}$. Evaluate the area given $a = 3.60$ cm, $b = 4.00$ cm and $c = 5.20$ cm

$$[A = 7.184 \text{ cm}^2]$$

12 Given that $a = 0.290$, $b = 14.86$, $c = 0.042$, $d = 31.8$ and $e = 0.650$ evaluate $v$ given that

$$v = \sqrt{\left(\frac{ab}{c} - \frac{d}{e}\right)}$$

$$[v = 7.327]$$

*Transposition of formulae*

Make the symbol indicated the subject of each of the formulae shown in *Problems 13 to 40*, and express each in its simplest form.

13 $a+b = c-d-e$ $\quad\quad$ $(d)$ $\quad\quad\quad\quad$ $[d = c-e-a-b]$

14 $x+3y = t$ $\quad\quad$ $(y)$ $\quad\quad\quad\quad$ $\left[y = \frac{1}{3}(t-x)\right]$

15 $c = 2\pi r$ $\quad\quad$ $(r)$ $\quad\quad\quad\quad$ $\left[r = \frac{c}{2\pi}\right]$

16 $y = mx+c$ $\quad\quad$ $(x)$ $\quad\quad\quad\quad$ $\left[x = \frac{y-c}{m}\right]$

17 $I = PRT$ $\quad\quad$ $(T)$ $\quad\quad\quad\quad$ $\left[T = \frac{I}{PR}\right]$

18 $I = \frac{E}{R}$ $\quad\quad$ $(R)$ $\quad\quad\quad\quad$ $\left[R = \frac{E}{I}\right]$

19 $S = \frac{a}{1-r}$ $\quad\quad$ $(r)$ $\quad\quad\quad\quad$ $\left[r = \frac{S-a}{S}\right]$

20 $F = \frac{9}{5}C+32$ $\quad\quad$ $(C)$ $\quad\quad\quad\quad$ $\left[C = \frac{5}{9}(F-32)\right]$

21 $y = \frac{\lambda(x-d)}{d}$ $\quad\quad$ $(x)$ $\quad\quad\quad\quad$ $\left[x = \frac{d}{\lambda}(y+\lambda)\right]$

22 $A = \frac{3(F-f)}{L}$ $\quad\quad$ $(f)$ $\quad\quad\quad\quad$ $\left[f = \frac{3F-AL}{3}\right]$

23 $y = \frac{Ml^2}{8EI}$ $\quad\quad$ $(E)$ $\quad\quad\quad\quad$ $\left[E = \frac{Ml^2}{8yI}\right]$

24 $R = R_0(1+\alpha t)$ $\quad\quad$ $(t)$ $\quad\quad\quad\quad$ $\left[t = \frac{R-R_0}{R_0\alpha}\right]$

25 $\frac{1}{R} = \frac{1}{R_1} + \frac{1}{R_2}$ $\quad\quad$ $(R_2)$ $\quad\quad\quad\quad$ $\left[R_2 = \frac{RR_1}{R_1-R}\right]$

26 $I = \frac{E-e}{R+r}$ $\quad\quad$ $(R)$ $\quad\quad\quad\quad$ $\left[R = \frac{E-e-Ir}{I}\right]$

27 $y = 4ab^2c^2$ $\quad\quad$ $(b)$ $\quad\quad\quad\quad$ $\left[b = \sqrt{\left(\frac{y}{4ac^2}\right)}\right]$

28 $\frac{a^2}{x^2} + \frac{b^2}{y^2} = 1$ $\quad\quad$ $(x)$ $\quad\quad\quad\quad$ $\left[x = \frac{ay}{\sqrt{(y^2-b^2)}}\right]$

29 $t = 2\pi\sqrt{\left(\frac{l}{g}\right)}$ $\quad\quad$ $(l)$ $\quad\quad\quad\quad$ $\left[l = \frac{t^2g}{4\pi^2}\right]$

30 $v^2 = u^2+2as$ $\quad\quad$ $(u)$ $\quad\quad\quad\quad$ $[u = \sqrt{(v^2-2as)}]$

31 $A = \dfrac{\pi R^2 \theta}{360}$          (R)          $\left[R = \sqrt{\left(\dfrac{360A}{\pi\theta}\right)}\right]$

32 $N = \sqrt{\left(\dfrac{a+x}{y}\right)}$          (a)          $[a = N^2 y - x]$

33 $Z = \sqrt{[R^2 + (2\pi fL)^2]}$          (L)          $\left[L = \dfrac{\sqrt{(Z^2 - R^2)}}{2\pi f}\right]$

34 $y = \dfrac{a^2 m - a^2 n}{x}$          (a)          $\left[a = \sqrt{\left(\dfrac{xy}{m-n}\right)}\right]$

35 $M = \pi(R^4 - r^4)$          (R)          $\left[R = \sqrt[4]{\left(\dfrac{M}{\pi} + r^4\right)}\right]$

36 $x + y = \dfrac{r}{3+r}$          (r)          $\left[r = \dfrac{3(x+y)}{(1-x-y)}\right]$

37 $m = \dfrac{\mu L}{L + rCR}$          (L)          $\left[L = \dfrac{mrCR}{\mu - m}\right]$

38 $a^2 = \dfrac{b^2 - c^2}{b^2}$          (b)          $\left[b = \dfrac{c}{\sqrt{(1-a^2)}}\right]$

39 $\dfrac{x}{y} = \dfrac{1+r^2}{1-r^2}$          (r)          $\left[r = \sqrt{\left(\dfrac{x-y}{x+y}\right)}\right]$

40 $\dfrac{p}{q} = \sqrt{\left(\dfrac{a+2b}{a-2b}\right)}$          (b)          $\left[b = \dfrac{a(p^2 - q^2)}{2(p^2 + q^2)}\right]$

41 A formula for the focal length, $f$, of a convex lens is $\dfrac{1}{f} = \dfrac{1}{u} + \dfrac{1}{v}$.
Transpose the formula to make $v$ the subject and evaluate $v$ when $f = 5$ and $u = 6$.
$$\left[v = \dfrac{uf}{u-f} \; ; 30\right]$$

42 The quantity of heat, $Q$, is given by the formula $Q = mc(t_2 - t_1)$. Make $t_2$ the subject of the formula and evaluate $t_2$ when $m = 10$, $t_1 = 15$, $c = 4$ and $Q = 1600$
$$\left[t_2 = t_1 + \dfrac{Q}{mc} \; ; 55\right]$$

43 The velocity, $v$, of water in a pipe appears in the formula
$h = \dfrac{0.03Lv^2}{2dg}$
Express $v$ as the subject of the formula and evaluate $v$ when $h = 0.712$, $L = 150$, $d = 0.30$ and $g = 9.81$
$$\left[v = \sqrt{\left(\dfrac{2dgh}{0.03L}\right)} \; ; 0.965\right]$$

44 The sag $S$ at the centre of a wire is given by the formula
$S = \sqrt{\left\{\dfrac{3d(l-d)}{8}\right\}}$
Make $l$ the subject of the formula and evaluate $l$ when $d = 1.75$ and $S = 0.80$
$$\left[l = \dfrac{8S^2}{3d} + d; 2.725\right]$$

45 In an electrical alternating current circuit the impedance $Z$ is given by
$$Z = \sqrt{\left\{R^2 + \left(\omega L - \dfrac{1}{\omega C}\right)^2\right\}}$$
Transpose the formula to make $C$ the subject and hence evaluate $C$ when $Z = 130$, $R = 120$, $\omega = 314$ and $L = 0.32$
$$\left[C = \dfrac{1}{\omega\{\omega L - \sqrt{(Z^2 - R^2)}\}} \; ; 63.1 \times 10^{-6}\right]$$

# 8 Straight line graphs

## A. MAIN POINTS CONCERNED WITH STRAIGHT LINE GRAPHS

1   The most common method of showing a relationship between two related sets of data is to use the **cartesian or rectangular axes**, as shown in *Fig 1*.

2   The points on a graph are called **co-ordinates**. Point A in figure 1 has the co-ordinates (3,2), i.e. $+3$ units in the $x$ direction and $+2$ units in the $y$ direction. Similarly, point B has co-ordinates $(-4,3)$ and C has co-ordinates $(-3,-2)$. The origin has co-ordinates (0,0).

3   The horizontal distance of a point from the vertical axis is called the **abscissa** and the vertical distance from the horizontal axis is called the **ordinate**.

4   The **gradient or slope** of a straight line is the ratio of the change in the value of $y$ to the change in the value of $x$ between any two points on the line. If, as $x$ increases ($\rightarrow$), $y$ also increases ($\uparrow$), then the gradient is positive. In *Fig 2(a)*, the gradient of AC is given by:

$$AC = \frac{\text{change in } y}{\text{change in } x} = \frac{CB}{BA} = \frac{7-3}{3-1} = \frac{4}{2} = 2$$

If as $x$ increases ($\rightarrow$), $y$ decreases ($\downarrow$), then the gradient is negative. In *Fig 2(b)*, the gradient of DF is given by:

$$DF = \frac{\text{change in } y}{\text{change in } x} = \frac{FE}{ED} = \frac{11-2}{-3-0} = \frac{9}{-3} = -3$$

5   (a) The process of finding equivalent values in between the given information is called **interpolation**.
    (b) The process of finding equivalent values which are outside of a given range of values is called **extrapolation** (see *Problem 1*).

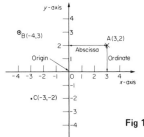

**Fig 1**

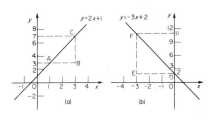

**Fig 2**

6 **Summary of general rules to be applied when drawing graphs**
   (i) Give the graph a title clearly explaining what is being illustrated.
   (ii) Choose scales such that the graph occupies as much space as possible on the graph paper being used.
   (iii) Choose scales so that interpolation is made as easy as possible. Usually, scales such as 1 cm = 1 unit, or 1 cm = 2 units, or 1 cm = 10 units are used. Awkward scales such as 1 cm = 3 units or 1 cm = 7 units are not used.
   (iv) The scales need not start at zero, if starting at zero produces an accumulation of points within a small area of the graph paper (see *Problem 2*).
   (v) The co-ordinates, or points, should be clearly marked. This may be done either by a cross, or by a dot and circle, or just by a dot (see *Fig 1*).
   (vi) A statement should be made next to each axis explaining the numbers represented with their appropriate units.
   (vii) Sufficient numbers should be written next to each axis without cramping.
   (viii) If the equation of a graph is of the form $y = mx + c$ the graph will always be a straight line, where $m$ represents the slope or gradient and $c$ the intercept with the $y$-axis, provided false zeros have not been introduced. With experimental results, do not join each point in turn but draw the best straight line through the co-ordinates.

## B. WORKED PROBLEMS ON STRAIGHT LINE GRAPHS

*Problem 1* The temperature in degrees Celsius and the corresponding values in degrees Fahrenheit are shown in the table below. Construct rectangular axes, choose a suitable scale and plot a graph of degrees Celsius (on the horizontal axis) against degrees Fahrenheit (on the vertical scale).

| °C | 10 | 20 | 40 | 60 | 80 | 100 |
|----|----|----|----|----|----|-----|
| °F | 50 | 68 | 104 | 140 | 176 | 212 |

From the graph find (a) the temperature in degrees Fahrenheit at 55°C, (b) the temperature in degrees Celsius at 167°F, (c) the Fahrenheit temperature at 0°C, and (d) the Celsius temperature at 230°F.

The co-ordinates (10,50), (20,68), (40,104), and so on are plotted as shown in *Fig 3*. When the co-ordinates are joined, a straight line is produced. Since a straight line results there is a linear relationship between degrees Celsius and degrees Fahrenheit.
(a) To find the Fahrenheit temperature at 55°C a vertical line AB is constructed from the horizontal axis to meet the straight line at B. The point where the horizontal line BD meets the vertical axis indicates the equivalent Fahrenheit temperature.
**Hence 55°C is equivalent to 131°F.**
This process of finding an equivalent value in between the given information in the above table is called **interpolation**.

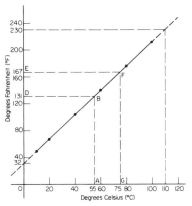

**Fig 3  Graph of degrees Celsius against degrees Fahrenheit**

(b) To find the Celsius temperature at 167°F, a horizontal line EF is constructed as shown in *Fig 3*. The point where the vertical line FG cuts the horizontal axis indicates the equivalent Celsius temperature. **Hence 167°F is equivalent to 75°C.**

(c) If the graph is assumed to be linear even outside of the given data, then the graph may be extended at both ends (shown by broken lines in *Fig 3*).

From *Fig 3*, 0°C corresponds to 32°F.

(d) **230°F is seen to correspond to 110°C.**

The process of finding equivalent values outside of the given range is called **extrapolation**.

*Problem 2*  In an experiment on Charles' law, the value of the volume of gas, $V$ m$^3$, was measured for various temperatures $T$°C. Results are shown below.

| $V$ m$^3$ | 25.0 | 25.8 | 26.6 | 27.4 | 28.2 | 29.0 |
|---|---|---|---|---|---|---|
| $T$°C | 60 | 65 | 70 | 75 | 80 | 85 |

Plot a graph of volume (vertical) against temperature (horizontal) and from it find (a) the temperature when the volume is 28.6 m$^3$ and (b) the volume when the temperature is 67°C.

If a graph is plotted with both the scales starting at zero then the result is as shown in *Fig 4*. All of the points lie in the top right-hand corner of the graph, making interpolation difficult. A more accurate graph is obtained if the temperature axis starts at 55°C and the volume axis starts at 24.5 m$^3$. The axes corresponding to these values is shown by the broken lines in *Fig. 4* and are called **false axes**, since the origin is not now at zero. A magnified version of this relevant part of the graph is shown in *Fig 5*. From the graph:

(a) When the volume is 28.6 m$^3$, the equivalent temperature is **82.5°C**, and
(b) When the temperature is 67°C, the equivalent volume is **26.1 m$^3$**.

*Problem 3*  Plot the graph $y = 4x + 3$ in the range $x = -3$ to $x = \pm 4$. From the graph, find (a) the value of $y$ when $x = 2.2$ and (b) the value of $x$ when $y = -3$

103

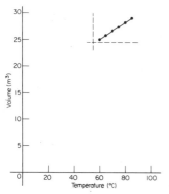

**Fig 4 (above) Graph of volume against temperature with a zero origin**

**Fig 5 (right) Graph of volume against temperature with a non-zero origin**

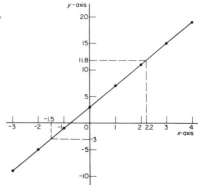

Whenever an equation is given and a graph is required, a table giving corresponding values of the variable is necessary. The table is achieved as follows:

When $x = -3$, $y = 4x+3 = 4(-3)+3 = -12+3 = -9$
When $x = -2$, $y = 4(-2)+3 = -8+3 = -5$, and so on

Such a table is shown below:

| $x$ | $-3$ | $-2$ | $-1$ | 0 | 1 | 2 | 3 | 4 |
|---|---|---|---|---|---|---|---|---|
| $y$ | $-9$ | $-5$ | $-1$ | 3 | 7 | 11 | 15 | 19 |

The co-ordinates $(-3,-9)$, $(-2,-5)$, $(-1,-1)$, and so on, are plotted and joined together to produce the straight line shown in *Fig 6*. (Note that the scales used on the $x$ and $y$ axes do not have to be the same.) From the graph:

(a) when $x = 2.2$, $y = 11.8$, and
(b) when $y = -3$, $x = -1.5$

**Fig 6  Graph of $y = 4x+3$**

104

**Problem 4** Plot the following graphs on the same axes between the range $x = -4$ to $x = +4$, and determine the gradients of each.
(a) $y = x$; (b) $y = x+2$; (c) $y = x+5$; (d) $y = x-3$

A table of co-ordinates is produced for each graph.

(a) $y = x$

| $x$ | $-4$ | $-3$ | $-2$ | $-1$ | 0 | 1 | 2 | 3 | 4 |
|-----|------|------|------|------|---|---|---|---|---|
| $y$ | $-4$ | $-3$ | $-2$ | $-1$ | 0 | 1 | 2 | 3 | 4 |

(b) $y = x+2$

| $x$ | $-4$ | $-3$ | $-2$ | $-1$ | 0 | 1 | 2 | 3 | 4 |
|-----|------|------|------|------|---|---|---|---|---|
| $y$ | $-2$ | $-1$ | 0 | 1 | 2 | 3 | 4 | 5 | 6 |

(c) $y = x+5$

| $x$ | $-4$ | $-3$ | $-2$ | $-1$ | 0 | 1 | 2 | 3 | 4 |
|-----|------|------|------|------|---|---|---|---|---|
| $y$ | 1 | 2 | 3 | 4 | 5 | 6 | 7 | 8 | 9 |

(d) $y = x-3$

| $x$ | $-4$ | $-3$ | $-2$ | $-1$ | 0 | 1 | 2 | 3 | 4 |
|-----|------|------|------|------|---|---|---|---|---|
| $y$ | $-7$ | $-6$ | $-5$ | $-4$ | $-3$ | $-2$ | $-1$ | 0 | 1 |

The co-ordinates are plotted and joined for each graph. The results are shown in *Fig 7*. Each of the straight lines produced are parallel to each other, i.e. the slope or gradient is the same for each.

To find the gradient of any straight line, say, $y = x-3$, a horizontal and vertical component needs to be constructed. In *Fig 7*, AB is constructed vertically at $x = 4$ and BC constructed horizontally at $y = -3$. The gradient of

$$AC = \frac{AB}{BC} = \frac{1-(-3)}{4-0} = \frac{4}{4} = 1$$

i.e. the gradient of the straight line $y = x-3$ is 1. The actual positioning of AB and BC is unimportant for the gradient is also given by

$$\frac{DE}{EF} = \frac{-1-(-2)}{2-1} = \frac{1}{1} = 1$$

**The slope or gradient of each of the straight lines in *Fig 7* is thus 1 since they are all parallel to each other.**

**Problem 5** Plot the following graphs on the same axes between the values $x = -3$ to $x = +3$ and determine the gradient and $y$-axis intercept of each.
(a) $y = 3x$; (b) $y = 3x+7$; (c) $y = -4x+4$; (d) $y = -4x-5$

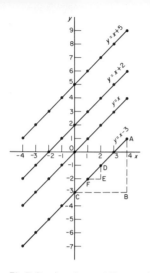

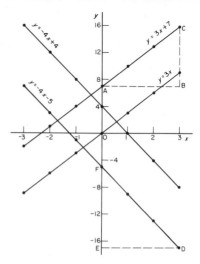

**Fig 7** Graphs of $y = x+5$, $y = x+2$, $y = x$ and $y = x-3$

**Fig 8** Graphs of $y = 3x$, $y = 3x+7$, $y = -4x+4$, $y = -4x-5$

A table of co-ordinates is drawn up for each equation.

(a) $y = 3x$

| $x$ | −3 | −2 | −1 | 0 | 1 | 2 | 3 |
|---|---|---|---|---|---|---|---|
| $y$ | −9 | −6 | −3 | 0 | 3 | 6 | 9 |

(b) $y = 3x+7$

| $x$ | −3 | −2 | −1 | 0 | 1 | 2 | 3 |
|---|---|---|---|---|---|---|---|
| $y$ | −2 | 1 | 4 | 7 | 10 | 13 | 16 |

(c) $y = -4x+4$

| $x$ | −3 | −2 | −1 | 0 | 1 | 2 | 3 |
|---|---|---|---|---|---|---|---|
| $y$ | 16 | 12 | 8 | 4 | 0 | −4 | −8 |

(d) $y = -4x-5$

| $x$ | −3 | −2 | −1 | 0 | 1 | 2 | 3 |
|---|---|---|---|---|---|---|---|
| $y$ | 7 | 3 | −1 | −5 | −9 | −13 | −17 |

Each of the graphs are plotted as shown in *Fig 8*, and each are straight lines. $y = 3x$ and $y = 3x+7$ are parallel to each other and thus have the same gradient.

Gradient of AC $= \dfrac{BC}{AB} = \dfrac{16-7}{3-0} = \dfrac{9}{3} = 3$

**Hence the gradient of both $y = 3x$ and $y = 3x+7$ is 3.**

$y = -4x+4$ and $y = -4x-5$ are parallel to each other and thus have the same gradient. The gradient of DF is given by

$$DF = \frac{EF}{ED} = \frac{-5-(-17)}{0-3} = \frac{12}{-3} = -4$$

**Hence the gradient of both $y = -4x+4$ and $y = -4x-5$ is $-4$**

The $y$-axis intercept means the value of $y$ where the straight line cuts the $y$-axis.
From *Fig 8*, $y = 3x$ cuts the $y$-axis at $y = 0$

$$y = 3x+7 \text{ cuts the } y\text{-axis at } y = +7$$
$$y = -4x+4 \text{ cuts the } y\text{-axis at } y = +4$$
$$\text{and } y = -4x-5 \text{ cuts the } y\text{-axis at } y = -5$$

Some general conclusions can be drawn from the graphs shown in *Figs 6, 7 and 8*. When an equation is of the form $y = mx+c$, where $m$ and $c$ are constants, then
(i)   a graph of $y$ against $x$ produces a straight line,
(ii)  $m$ represents the slope or gradient of the line, and
(iii) $c$ represents the $y$-axis intercept.
Thus, given an equation such as $y = 3x+7$, it may be deduced 'on sight' that its gradient is $+3$ and its $y$-axis intercept is $+7$, as shown in *Fig 8*. Similarly, if $y = -4x-5$, then the gradient is $-4$ and the $y$-axis intercept is $-5$, as shown in *Fig 8*.

When plotting a graph of the form $y = mx+c$, only two co-ordinates need be determined. When the co-ordinates are plotted a straight line is drawn between the two points. Normally, three co-ordinates are determined, the third one acting as a check.

*Problem 6* The following equations represent straight lines. Determine, without plotting graphs, the gradient and $y$-axis intercept for each.
(a) $y = 3$; (b) $y = 2x$; (c) $y = 5x-1$; (d) $2x+3y = 3$

(a) $y = 3$ (which is of the form $y = 0x+3$) represents a horizontal straight line intercepting the $y$-axis at 3. Since the line is horizontal its **gradient is zero.**
(b) $y = 2x$ is of the form $y = mx+c$, where $c$ is zero. Hence **gradient = 2** and **$y$-axis intercept = 0** (i.e., the origin).
(c) $y = 5x-1$ is of the form $y = mx+c$. Hence **gradient = 5** and **$y$-axis intercept = $-1$.**
(d) $2x+3y = 3$ is not in the form $y = mx+c$ as it stands. Transposing to make $y$ the subject gives $3y = 3-2x$

$$\text{i.e.} \quad y = \frac{3-2x}{3} = \frac{3}{3} - \frac{2x}{3}$$

$$\text{i.e.} \quad y = -\frac{2x}{3} + 1, \text{ which is of the form } y = mx+c$$

Hence **gradient = $-\frac{2}{3}$** and **$y$-axis intercept = $+1$**

*Problem 7* In an experiment demonstrating Hooke's law, the strain in an aluminium wire was measured for various stresses. The results were:

| Stress N/mm$^2$ | 4.9 | 8.7 | 15.0 | 18.4 | 24.2 | 27.3 |
|---|---|---|---|---|---|---|
| Strain | 0.000 07 | 0.000 13 | 0.000 21 | 0.000 27 | 0.000 34 | 0.000 39 |

107

Plot a graph of stress (vertically) against strain (horizontally). Find:

(a) Young's Modulus of elasticity for aluminium, which is given by the gradient of the graph,

(b) the value of the strain at a stress of 20 N/mm², and

(c) the value of the stress when the strain is 0.000 20

The co-ordinates (0.000 07, 4.9), (0.000 13, 8.7), and so on, are plotted as shown in *Fig 9*. The graph produced is the best straight line which can be drawn corresponding to these points. (With experimental results it is unlikely that all the points will lie exactly on a straight line.) The graph, and each of its axes, are labelled. Since the straight line passes through the origin, then stress is directly proportional to strain for the given range of values.

(a) The gradient of the straight line,

$$AC = \frac{AB}{BC} = \frac{28-7}{0.000\ 40-0.000\ 10} = \frac{21}{0.000\ 30}$$

$$= \frac{21}{3 \times 10^{-4}} = \frac{7}{10^{-4}} = 7 \times 10^4$$

$$= 70\ 000\ \text{N/mm}^2$$

**Thus Young's Modulus of Elasticity for aluminium is 70 000 N/mm²**
Since $1\,\text{m}^2 = 10^6\ \text{mm}^2$, 70 000 N/mm² is equivalent to $70\ 000 \times 10^6$ N/m²
i.e., **$70 \times 10^9$ N/m² (or Pascals)**
From *Fig 9*:

(b) the value of the strain at a stress of 20N/mm² is **0.000 285**, and

(c) the value of the stress when the strain is 0.000 20 is **14 N/mm²**

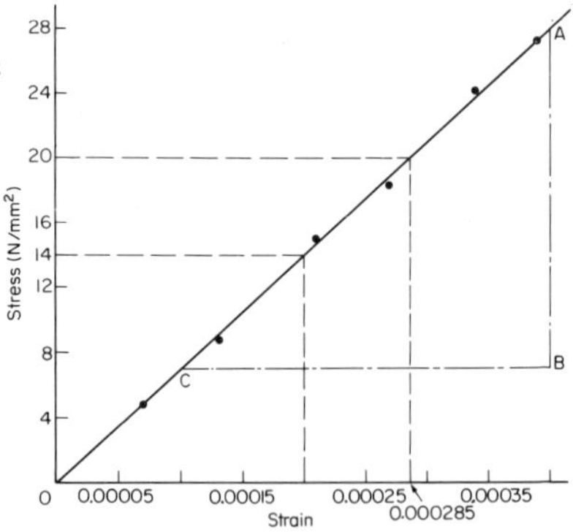

**Fig 9  Graph of stress against strain for aluminium**

Problem 8  The following values of resistance $R$ ohms and corresponding voltage $V$ volts are obtained from a test on a filament lamp.

| $R$ ohms | 30 | 48.5 | 73 | 107 | 128 |
|---|---|---|---|---|---|
| $V$ volts | 16 | 29 | 52 | 76 | 94 |

Choose suitable scales and plot a graph with $R$ representing the vertical axis and $V$ the horizontal axis. Determine (a) the slope of the graph, (b) the $R$ axis intercept value, (c) the equation of the graph, (d) the value of resistance when the voltage is 60 V, and (e) the value of the voltage when the resistance is 40 ohms. (f) If the graph were to continue in the same manner, what value of resistance would be obtained at 110 V?

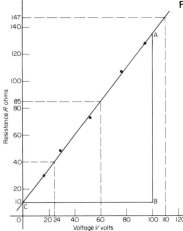

**Fig 10 Graph of resistance against voltage**

The co-ordinates (16,30), (29,48.5), and so on, are shown plotted in *Fig 10* where the best straight line is drawn through the points.

(a) The slope or gradient of the straight line,

$$AC = \frac{AB}{BC} = \frac{135-10}{100-0} = \frac{125}{100} = 1.25$$

(Note that the vertical line AB and the horizontal line BC may be constructed anywhere along the length of the straight line. However, calculations are made easier if the horizontal line BC is carefully chosen, in this case, 100.)

(b) The $R$ axis intercept is at $R = \mathbf{10}$ **ohms** (by extrapolation).

(c) The equation of a straight line is $y = mx+c$, when $y$ is plotted on the vertical axis and $x$ on the horizontal axis. $m$ represents the gradient and $c$ the $y$-axis intercept. In this case, $R$ corresponds to $y$, $V$ corresponds to $x$, $m = 1.25$ and $c = 10$. Hence the equation of the graph is $R = \mathbf{(1.25V + 10)}\ \Omega$.

From *Fig 10*,

(d) when the voltage is 60 V, the resistance is **85 $\Omega$**,

(e) when the resistance is 40 ohms, the voltage is **24 V**, and

(f) by extrapolation, when the voltage is 110 V, the resistance is **147 $\Omega$**

## C. FURTHER PROBLEMS ON STRAIGHT LINE GRAPHS

1   Corresponding values obtained experimentally for two quantities are:

| $x$ | −2.0 | −0.5 | 0 | 1.0 | 2.5 | 3.0 | 5.0 |
|---|---|---|---|---|---|---|---|
| $y$ | −13.0 | −5.5 | −3.0 | 2.0 | 9.5 | 12.0 | 22.0 |

Use a horizontal scale for $x$ of 1 cm = $\frac{1}{2}$ unit and a vertical scale for $y$ of 1 cm = 2 units and draw a graph of $x$ against $y$. Label the graph and each of its axes. By interpolation, find from the graph the value of $y$ when $x$ is 3.5.

[14.5]

2   The equation of a line is $4y = 2x+5$. A table of corresponding values is produced and is shown below. Complete the table and plot a graph of $y$ against $x$. Find the gradient of the graph.

| $x$ | −4 | −3 | −2 | −1 | 0 | 1 | 2 | 3 | 4 |
|---|---|---|---|---|---|---|---|---|---|
| $y$ | | −0.50 | | | 1.25 | | | | 3.25 |

$\left[\frac{1}{2}\right]$

3   Determine the gradient and intercept on the $y$-axis for each of the following equations:
(a) $y = 4x-2$; (b) $y = -x$; (c) $y = -3x-4$; (d) $y = 4$

$$[(a) 4,-2; (b) -1,0; (c) -3,-4; (d) 0,4]$$

4   Find the gradient and intercept on the $y$-axis for each of the following equations:
(a) $2y-1 = 4x$; (b) $6x-2y = 5$; (c) $3(2y-1) = x/4$

$$\left[(a)\ 2,\tfrac{1}{2};\ (b)\ 3,-2\tfrac{1}{2};\ (c)\ \tfrac{1}{24},\tfrac{1}{2}\right]$$

5   Draw on the same axes the graphs of $y = 3x-5$ and $3y+2x = 7$. Find the co-ordination of the point of intersection. Check the result obtained by solving the two simultaneous equations algebraically.

$$[(2, 1)]$$

6   The resistance $R$ ohms of a copper winding is measured at various temperatures $t°C$ and the results are as follows:

| $R$ ohms | 112 | 120 | 126 | 131 | 136 |
|----------|-----|-----|-----|-----|-----|
| $t°C$    | 20  | 36  | 48  | 58  | 64  |

Plot a graph of $R$ (vertically) against $t$ (horizontally) and find from it (a) the temperature when the resistance is 122 $\Omega$ and (b) the resistance when the temperature is 52°C.

$$[(a) 40°C; (b) 128 \Omega]$$

7   The speed of a motor varies with armature voltage as shown by the following experimental results:

| $n$ (rev/min) | 285 | 517 | 615 | 750 | 917 | 1050 |
|---------------|-----|-----|-----|-----|-----|------|
| $V$ (volts)   | 60  | 95  | 110 | 130 | 155 | 175  |

Plot a graph of speed (horizontally) against voltage (vertically) and draw the best straight line through the points. Find from the graph (a) the speed at a voltage of 145 V, and (b) the voltage at a speed of 400 rev/min.

$$[(a) 850 \text{ rev/min}; (b) 77.5 \text{ V}]$$

8   The following table gives the force $F$ newtons which, when applied to a lifting machine overcomes a corresponding load of $L$ newtons.

| Force $F$ newtons | 25 | 47 | 64 | 120 | 149 | 187 |
|-------------------|----|----|----|-----|-----|-----|
| Load $L$ newtons  | 50 | 140 | 210 | 430 | 550 | 700 |

Choose suitable scales and plot a graph of $F$ (vertically) against $L$ (horizontally). Draw the best straight line through the points. Determine from the graph (a) the gradient, (b) the $F$-axis intercept, (c) the equation of the graph, (d) the force applied when the load is 310 N, and (e) the load that a force of 160 N will overcome. (f) If the graph were to continue in the same manner, what value of force will be needed to overcome a 800 N load?

$$\left[\begin{array}{l}(a)\ 0.25;\ (b)\ 12;\ (c)\ F = 0.25L+12; \\ (d)\ 89.5\ \text{N};\ (e)\ 592\ \text{N};\ (f)\ 212\ \text{N}\end{array}\right]$$

9 The following table gives the results of tests carried out to determine the breaking stress $\sigma$ of rolled copper at various temperatures, $t$:

| Stress $\sigma$ (N/cm²) | 8.51 | 8.07 | 7.80 | 7.47 | 7.23 | 6.78 |
|---|---|---|---|---|---|---|
| Temperature $t$ (°C) | 75 | 220 | 310 | 420 | 500 | 650 |

Plot a graph of stress (vertically) against temperature (horizontally). Draw the best straight line through the plotted co-ordinates. Determine the slope of the graph and the vertical axis intercept.

$[-0.003; 8.73]$

10 The velocity $v$ of a body after varying time intervals $t$ was measured as follows:

| $t$ (seconds) | 2 | 5 | 8 | 11 | 15 | 18 |
|---|---|---|---|---|---|---|
| $v$ (m/s) | 16.9 | 19.0 | 21.1 | 23.2 | 26.0 | 28.1 |

Plot $v$ vertically and $t$ horizontally and draw a graph of velocity against time. Determine from the graph (a) the velocity after 10 s; (b) the time at 20 m/s and (c) the equation of the graph.

$[(a) 22.5 \text{ m/s}; (b) 6.43 \text{ s}; (c) v = 0.7t+15.5]$

11 The mass $m$ of a steel joist varies with length $l$ as follows:

| mass, $m$ (kg) | 80 | 100 | 120 | 140 | 160 |
|---|---|---|---|---|---|
| length, $l$ (m) | 3.00 | 3.74 | 4.48 | 5.23 | 5.97 |

Plot a graph of mass (vertically) against length (horizontally). Determine the equation of the graph.

$[m = 26.9l-0.63]$

12 The crushing strength of mortar varies with the percentage of water used in its preparation, as shown below.

| Crushing strength, $F$ (tonnes) | 1.64 | 1.36 | 1.07 | 0.78 | 0.50 | 0.22 |
|---|---|---|---|---|---|---|
| % of water used, $w$% | 6 | 9 | 12 | 15 | 18 | 21 |

Plot a graph of $F$ (vertically) against $w$ (horizontally).
(a) Interpolate and determine the crushing strength when 10% of water is used.
(b) Assuming the graph continues in the same manner extrapolate and determine the percentage of water used when the crushing strength is 0.15 tonnes.
(c) What is the equation of the graph?

$\begin{bmatrix} \text{(a) } 1.26 \text{ t; (b) } 21.68\% \\ \text{(c) } F = -0.095 \, w + 2.21 \end{bmatrix}$

# 9 Geometry

## A. MAIN POINTS CONCERNED WITH GEOMETRY

1  **Geometry** is a part of mathematics in which the properties of points, lines, surfaces and solids are investigated.

2  An **angle** is the amount of rotation between two straight lines. Angles may be measured in either **degrees** or **radians** (see paragraph 15). **1 revolution = 360 degrees**, thus 1 degree = $\frac{1}{360}$ th of one revolution. Also 1 minute = $\frac{1}{60}$ th of a degree and 1 second = $\frac{1}{60}$ th of a minute. 1 minute is written as 1' and 1 second is written as 1''. **Thus 1° = 60' and 1' = 60''**.

3  (i)   Any angle between 0° and 90° is called an **acute angle**.
   (ii)  An angle equal to 90° is called a **right angle**.
   (iii) Any angle between 90° and 180° is called an **obtuse angle**.
   (iv)  Any angle greater than 180° and less than 360° is called a **reflex angle**.

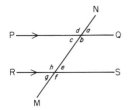

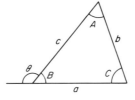

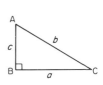

**Fig 1**          **Fig 2**          **Fig 3**

4  (i)   An angle of 180° lies on a straight line.
   (ii)  If two angles add up to 90° they are called **complementary angles**.
   (iii) If two angles add up to 180° they are called **supplementary angles**.
   (iv)  **Parallel lines** are straight lines which are in the same plane and never meet. (Such lines are denoted by arrows, as in *Fig 1*.)
   (v)   A straight line which crosses two parallel lines is called a **transversal** (see MN in *Fig 1*).

5  With reference to *Fig 1*:
   (i)   $a = c$, $b = d$, $e = g$ and $f = h$. Such pairs of angles are called **vertically opposite angles**.
   (ii)  $a = e$, $b = f$, $c = g$ and $d = h$. Such pairs of angles are called **corresponding angles**.
   (iii) $c = e$ and $b = h$. Such pairs of angles are called **alternate angles**.
   (iv)  $b+e = 180°$ and $c+h = 180°$. Such pairs of angles are called **interior angles**.

6  A **triangle** is a figure enclosed by three straight lines. The sum of the three angles of a triangle is equal to $180°$.

7  **Types of triangles:**
   (i)  An **acute-angled triangle** is one in which all the angles are acute, i.e. all the angles are less than $90°$.
   (ii)  A **right-angled triangle** is one which contains a right angle.
   (iii)  An **obtuse-angled triangle** is one which contains an obtuse angle, i.e. one angle which lies between $90°$ and $180°$.
   (iv)  An **equilateral triangle** is one in which all the sides and all the angles are equal (i.e. each $60°$).
   (v)  An **isosceles triangle** is one in which two angles and two sides are equal.
   (vi)  A **scalene triangle** is one with unequal angles and therefore unequal sides.

8  With reference to *Fig 2*:
   (i)  Angles $A$, $B$ and $C$ are called **interior angles** of the triangle.
   (ii)  Angle $\theta$ is called an **exterior angle** of the triangle and is equal to the sum of the two opposite interior angles, i.e. $\theta = A + C$.
   (iii)  $a+b+c$ is called the **perimeter** of the triangle.

9  With reference to *Fig 3*, the side opposite the right angle (side $b$) is called the **hypotenuse**. The **theorem of Pythagoras** states: 'In any right-angled triangle, the square on the hypotenuse is equal to the sum of the squares on the other two sides.' Hence $b^2 = a^2 + c^2$.

10  Two triangles are said to be **congruent** if they are equal in all respects, i.e. three angles and three sides in one triangle are equal to three angles and three sides in the other triangle. Two triangles are congruent if:
   (i)  the three sides of one are equal to the three sides of the other (SSS),
   (ii)  they have two sides of the one equal to two sides of the other, and if the angles included by three sides are equal (SAS),
   (iii)  two angles of the one are equal to two angles of the other and any side of the first is equal to the corresponding side of the other (ASA), or
   (iv)  their hypotenuses are equal and if one other side of one is equal to the corresponding side of the other (RHS).

11  Two triangles are said to be **similar** if the angles of one triangle are equal to the angles of the other triangle. With reference to *Fig 4*: Triangles ABC and PQR are similar and the corresponding sides are in proportion to each other, i.e.

$$\frac{p}{a} = \frac{q}{b} = \frac{r}{c}$$

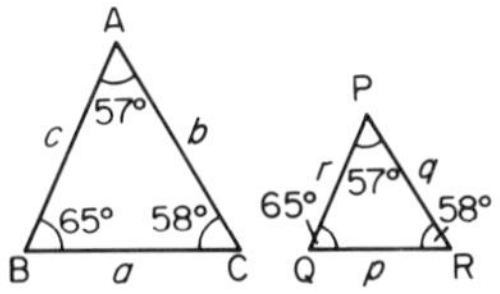

**Fig 4**

12  To **construct** any triangle the following drawing instruments are needed: (i) ruler and/or straight edge, (ii) compass, (iii) protractor, (iv) pencil. For actual constructions, see *Problems 26 to 29*.

13  A **circle** is a plain figure enclosed by a curved line, every point on which is equidistant from a point within, called the **centre**.

14  **Properties of circles:**
   (i)  The distance from the centre to the curve is called the **radius, $r$**, of the circle (see OP in *Fig 5*).
   (ii)  The boundary of a circle is called the **circumference, $c$**.
   (iii)  Any straight line passing through the centre and touching the circumference at each end is called the **diameter, $d$**, (see QR in *Fig 5*). Thus $d = 2r$.

113

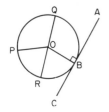

**Fig 5**      **Fig 6**      **Fig 7**      **Fig 8**

(iv) The ratio

$$\frac{\text{circumference}}{\text{diameter}} = \text{a constant for any circle.}$$

This constant is denoted by the Greek letter $\pi$ (pronounced 'pie'), where $\pi = 3.14159$, correct to 5 decimal places.
Hence $c/d = \pi$ or $c = \pi d$ or $c = 2\pi r$.

(v) A **semicircle** is one half of the whole circle.

(vi) A **quadrant** is one quarter of a whole circle.

(vii) A **tangent** to a circle is a straight line which meets the circle in one point only and does not cut the circle when produced. AC in *Fig 5* is a tangent to the circle since it touches the curve at point B only. If radius OB is drawn, then angle ABO is a right angle.

(viii) A **sector** of a circle is the part of a circle between radii (for example, the portion OXY of *Fig 6* is a sector). If a sector is less than a semicircle it is called a **minor sector**, if greater than a semicircle it is called a **major sector**.

(ix) A **chord** of a circle is any straight line which divides the circle into two parts and is terminated at each end by the circumference. ST, in *Fig 6*, is a chord.

(x) A **segment** is the name given to the parts into which a circle is divided by a chord. If the segment is less than a semi-circle it is called a **minor segment** (see shaded area in *Fig 6*). If the segment is greater than a semicircle it is called a **major segment** (see the unshaded area in *Fig 6*).

(xi) An **arc** is a portion of the circumference of a circle. The distance SRT in *Fig 6* is called a **minor arc** and the distance SXYT is called a **major arc**.

(xii) The angle at the centre of a circle, subtended by an arc, is double the angle at the circumference subtended by the same arc. With reference to *Fig 7*, **Angle AOC = 2 × angle ABC**.

(xiii) The angle in a semicircle is a right angle (see angle BQP in *Fig 7*).

15 One **radian** is defined as the angle subtended at the centre of a circle by an arc equal in length to the radius. With reference to *Fig 8*, for arc length $l$, $\theta$ radians $= l/r$ or $l = r\theta$, where $\theta$ is in radians. When $l = $ whole circumference $(= 2\pi r)$ then $\theta = l/r = 2\pi r/r = 2\pi$,
i.e. **$2\pi$ radians = $360°$** or **$\pi$ radians = $180°$**.

Thus $1 \text{ rad} = \dfrac{180°}{\pi} = 57.30°$, correct to 2 decimal places.

Since $\pi$ rads $= 180°$, then $\dfrac{\pi}{2}$ rads $= 90°$, $\dfrac{\pi}{3}$ rads $= 60°$, $\dfrac{\pi}{4}$ rads $= 45°$, and so on.

## B. WORKED PROBLEMS ON GEOMETRY

### (a) ANGULAR MEASUREMENT

*Problem 1.* Add 14° 53′ and 37° 18′

|            |                                                                      |
| ---------- | -------------------------------------------------------------------- |
| 14° 53′    | 53′+19′ = 72′. Since 60′ = 1°, 72′ = 1° 12′.                          |
| 37° 19′    | Thus the 12′ is placed in the minutes column and 1° is carried in    |
| 52° 12′    | the degrees column. Then 14°+37°+1° (carried) = 52°.                 |
| 1°         |                                                                      |

Thus 14° 53′+37° 19′ = 52° 12′

*Problem 2* Subtract 15° 47′ from 28° 13′

27°
28° 13′      13′−47′ cannot be done. Hence 1° or 60′ is 'borrowed' from the
15° 47′      degrees column, which leaves 27° in that column. Now
12° 26′      (60′+13′)−47′ = 26′, which is placed in the minutes column.
                 27°−15° = 12°, which is placed in the degrees column.

Thus 28° 13′−15° 47′ = 12° 26′

*Problem 3* Determine (a) 13° 42′ 51″+48° 22′ 17″, (b) 37° 12′ 8″−21° 17′ 25″

(a)     13° 42′ 51″          (b)                36° 11′
           48° 22′ 17″                   37° 12′ 8″
Adding:   62° 5′ 8″           Subtracting:   21° 17′ 25″
            1° 1′                          15° 54′ 43″

*Problem 4* Convert (a) 24° 42′, (b) 78° 15′ 26″ to degrees and decimals of a degree

(a) Since 1 minute = $\frac{1}{60}$th of a degree, $42′ = \frac{42°}{60} = 0.70°$

    **Hence 24° 42′ = 24.70°**

(b) Since 1 second = $\frac{1}{60}$th of a minute, $26″ = \frac{26′}{60} = 0.4333′$

    Hence 78° 15′ 26″ = 78° 15.4333′

    $15.4333′ = \frac{15.4333°}{60} = 0.2572°$, correct to 4 decimal places.

    **Hence 78° 15′ 26″ = 78.2572°, correct to 4 decimal places**

**Problem 5** Convert 45.371° into degrees, minutes and seconds

Since $1° = 60'$, $0.371° = (0.371 \times 60)' = 22.26'$.
Since $1' = 60''$, $0.26' = (0.26 \times 60)'' = 15.6'' = 16''$ to the nearest second
**Hence 45.371° = 45° 22' 16''**

*Further problems on angular measurement may be found in section C, problems 1 to 4, page 127.*

## (b) TYPES AND PROPERTIES OF ANGLES

**Problem 6** State the general name given to the following angles:
(a) 159°; (b) 63°; (c) 90°; (d) 227°

(a) 159° lies between 90° and 180° and is therefore called an **obtuse angle.**
(b) 63° lies between 0° and 90° and is therefore called an **acute angle.**
(c) 90° is called a **right angle.**
(d) 227° is greater than 180° and less than 360° and is therefore called a **reflex angle.**

**Problem 7** Find the angles complementary to (a) 41°, (b) 58° 39'

(a) The complement of 41° is (90° −41°), i.e. **49°**
(b) The complement of 58° 39' is (90° −58° 39'), i.e. **31° 21'**

**Problem 8** Find the angles supplementary to (a) 27°; (b) 111° 11'

(a) The supplement of 27° is (180° −27°), i.e. **153°**
(b) The supplement of 111° 11' is (180° −111° 11'), i.e. **68° 49'**

**Problem 9** Two straight lines AB and CD intersect at 0. If ∠AOC is 43°, find ∠AOD, ∠DOB and ∠BOC

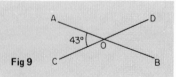

**Fig 9**

From *Fig 9*, ∠AOD is supplementary to ∠AOC. Hence ∠AOD = 180° −43° = 137°. When two straight lines intersect the vertically opposite angles are equal. Hence ∠**DOB** = **43°** and ∠**BOC** = **137°**

116

α = 180°−133° = 47° (i.e. supplementary angles).
α = β = **47°** (corresponding angles between parallel lines)

*Problem 11* Determine the value of angle θ in *Fig 11*

Let a straight line FG be drawn through E such that FG is parallel to AB and CD.
∠BAE = ∠AEF (alternate angles between parallel lines AB and FG).
Hence ∠AEF = 23° 37′
∠ECD = ∠FEC (alternate angles between parallel lines FG and CD).
Hence ∠FEC = 35° 49′
Angle θ = ∠AEF+∠FEC = 23° 37′+35° 49′ = **59° 26′**

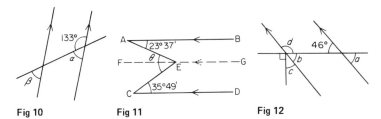

**Fig 10**          **Fig 11**          **Fig 12**

*Problem 12* Determine angles *c* and *d* in *Fig 12*

*b* = 46° (corresponding angles between parallel lines)
Also *b*+*c*+90° = 180° (angles on a straight line)
Hence 46°+*c*+90° = 180°, from which *c* = **44°**
*b* and *d* are supplementary, hence *d* = 180°−46° = **134°**
[Alternatively, 90°+*c* = *d* (vertically opposite angles)]

*Further problems on types and properties of angles may be found in section C, problems 5 to 11, page 128.*

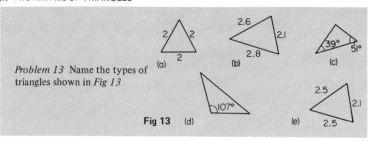

*Problem 13* Name the types of triangles shown in *Fig 13*

**Fig 13**

(a) Equilateral triangle.
(b) Acute-angled scalene triangle.
(c) Right-angled triangle.

(d) Obtuse-angled scalene triangle.
(e) Isosceles triangle.

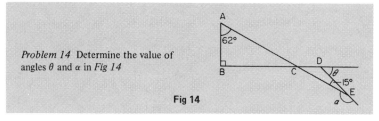

*Problem 14* Determine the value of angles $\theta$ and $\alpha$ in *Fig 14*

**Fig 14**

In triangle ABC, $\angle A + \angle B + \angle C = 180°$. (Angles in a triangle add up to $180°$)
Hence $\angle C = 180° - 90° - 62° = 28°$
Thus $\angle DCE = 28°$ (vertically opposite angles)
$\angle\theta = \angle DCE + \angle DEC$ (exterior angle of a triangle is equal to the sum of the two opposite interior angles)
Hence $\angle\theta = 28° + 15° = \mathbf{43°}$
$\angle\alpha$ and $\angle DEC$ are supplementary, thus $\alpha = 180° - 15° = \mathbf{165°}$

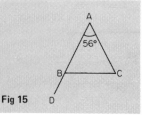

*Problem 15* ABC is an isosceles triangle in which the unequal angle BAC is $56°$. AB is extended to D as shown in *Fig 15*. Determine the angle DBC

**Fig 15**

Since the three interior angles of a triangle add up to $180°$ then
$56° + \angle B + \angle C = 180°$, i.e. $\angle B + \angle C = 180° - 56° = 124°$
Triangle ABC is isosceles hence $\angle B = \angle C = \dfrac{124°}{2} = 62°$
$\angle DBC = \angle A + \angle C$ (exterior angle equals sum of two interior opposite angles),
i.e. $\angle DBC = 56° + 62° = \mathbf{118°}$
[Alternatively, $\angle DBC + \angle ABC = 180°$ (i.e. supplementary angles)]

**Fig 16**

*Problem 16* Find angles *a*, *b*, *c*, *d* and *e* in *Fig 16*

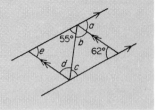

$a = 62°$ and $c = 55°$ (alternate angles between parallel lines)
$55° + b + 62° = 180°$ (angles in a triangle add up to 180°)
Hence $b = 180° - 55° - 62° = 63°$
$b = d = 63°$ (alternate angles between parallel lines)
$e + 55° + 63° = 180°$ (angles in a triangle add up to 180°)
Hence $e = 180° - 55° - 63° = 62°$
[Check: $e = a = 62°$ (corresponding angles between parallel lines)]

*Further problems on properties of triangles may be found in section C, problems 12 to 16, page 129.*

(d) THEOREM OF PYTHAGORAS

*Problem 17* In *Fig 17*, find the length of BC

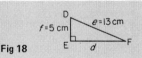

**Fig 17**

By Pythagoras' theorem: $a^2 = b^2 + c^2$
i.e. $a^2 = 4^2 + 3^2 = 16 + 9 = 25$
Hence $a = \sqrt{25} = \pm 5$ (−5 has no meaning in this context and is thus ignored)

**Thus BC = 5 cm**

*Problem 18* In *Fig 18*, find the length of EF

**Fig 18**

By Pythagoras' theorem: $e^2 = d^2 + f^2$
Hence $13^2 = d^2 + 5^2$
$169 = d^2 + 25$
$d^2 = 169 - 25 = 144$
Thus $d = \sqrt{144} = 12$ cm
**Thus EF = 12 cm**

**Problem 19** Two aircraft leave an airfield at the same time. One travels due north at an average speed of 300 km/h and the other due west at an average speed of 220 km/h. Calculate their distance apart after 4 hours.

After 4 hours, the first aircraft has travelled 4 × 300 = 1 200 km, due north, and the second aircraft has travelled 4 × 220 = 880 km due west, as shown in *Fig 19*.

Distance apart after 4 hours = BC.

From Pythagoras' theorem:

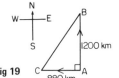

$$BC^2 = 1\ 200^2 + 880^2$$
$$= 1\ 440\ 000 + 774\ 400 \quad \text{and} \quad BC = \sqrt{(2\ 214\ 400)}$$

**Hence distance apart after 4 hours = 1 488 km**

**Fig 19**

*Further problems on the theorem of Pythagoras may be found in section C, Problems 17 to 23, page 129.*

(e) CONGRUENT TRIANGLES

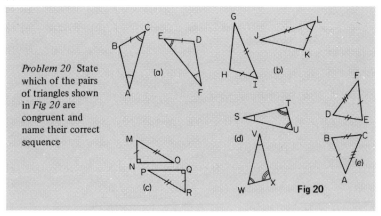

**Problem 20** State which of the pairs of triangles shown in *Fig 20* are congruent and name their correct sequence

**Fig 20**

(a) Congruent ABC, FDE (Angle, side, angle, i.e. ASA).
(b) Congruent GIH, JLK (Side, angle, side, i.e. SAS).
(c) Congruent MNO, RQP (Right-angle, hypotenuse, side, i.e. RHS).
(d) Not necessarily congruent. It is not indicated that any side coincides.
(e) Congruent ABC, FED (Side, side, side, i.e. SSS).

**Problem 21** In *Fig 21*, triangle PQR is isosceles with Z the mid-point of PQ. Prove that triangle PXZ and QYZ are congruent, and that triangles RXZ and RYZ are congruent. Determine the values of angles RPZ and RXZ

**Fig 21**

Since triangle PQR is isosceles PR = RQ and thus $\angle QPR = \angle RQP$.

$\angle RXZ = \angle QPR + 28°$ and $\angle RYZ = \angle RQP + 28°$ (exterior angles of a triangle equals the sum of the two interior opposite angles).

Hence $\angle RXZ = \angle RYZ$.

$\angle PXZ = 180° - \angle RXZ$ and $\angle QYZ = 180° - \angle RYZ$. Thus $\angle PXZ = \angle QYZ$.

Triangles PXZ and QYZ are congruent since $\angle XPZ = \angle YQZ$, PZ = ZQ and $\angle XZR = \angle YZQ$ (ASA). Hence XZ = YZ

Triangles PRZ and QRZ are congruent since PR = RQ, $\angle RPZ = \angle RQZ$ and PZ = ZQ (SAS). Hence $\angle RZX = \angle RZY$

Triangles RXZ and RYZ are congruent since $\angle RXZ = \angle RYZ$, XZ = YZ and $\angle RZX = \angle RZY$ (ASA)

$\angle QRZ = 67°$ and thus $\angle PRQ = 67° + 67° = 134°$

Hence $\quad \angle RPZ = \angle RQZ = \dfrac{180° - 134°}{2} = 23°$

$\angle RXZ = 23° + 28° = 51°$ (external angle of a triangle equals the sum of the two interior opposite angles).

*Further problems on congruent triangles may be found in section C, problems 24 and 25, page 130.*

### (f) SIMILAR TRIANGLES

*Problem 22* In *Fig 22*, find the length of side $a$.

In triangle ABC, $50° + 30° + \angle C = 180°$, from which $\angle C = 60°$

In triangle DEF, $\angle E = 180° - 50° - 60° = 70°$

Hence triangles ABC and DEF are similar, since their angles are the same. Since corresponding sides are in proportion to each other then:

$\dfrac{a}{d} = \dfrac{c}{f}$. i.e. $\dfrac{a}{4.42} = \dfrac{12.0}{5.0}$

Hence $a = \dfrac{12.0}{5.0}(4.42) = $ **10.61 cm**

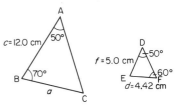

**Fig 22**

*Problem 23* In *Fig 23*, find the dimensions marked $p$, $r$ and $z$

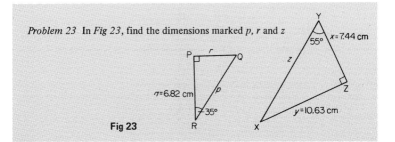

**Fig 23**

In triangle PQR, $\angle Q = 180° - 90° - 35° = 55°$
In triangle XYZ, $\angle X = 180° - 90° - 55° = 35°$

Hence triangles PQR and XYZ are similar since their angles are the same. The triangles may be redrawn as shown in *Fig 24*.

By proportion: $\dfrac{p}{z} = \dfrac{r}{x} = \dfrac{q}{y}$

Hence $\qquad \dfrac{p}{z} = \dfrac{r}{7.44} = \dfrac{6.82}{10.63}$

from which, $\qquad r = 7.44 \left(\dfrac{6.82}{10.63}\right) = \textbf{4.77 cm}$

Using Pythagoras' theorem on triangle XYZ gives:

$z = \sqrt{[(7.44)^2 + (10.63)^2]} = \textbf{12.97 cm}$

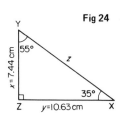

**Fig 24**

By proportion: $\dfrac{p}{z} = \dfrac{q}{y}$, i.e. $\dfrac{p}{12.97} = \dfrac{6.82}{10.63}$

Hence $\qquad p = 12.97 \left(\dfrac{6.82}{10.63}\right) = \textbf{8.32 cm}$

*Problem 24* In *Fig 25*, show that triangles CBD and CAE are similar and hence find the length of CD and BD

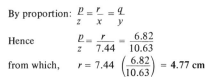

**Fig 25**

Since BD is parallel to AE then:

$\angle CBD = \angle CAE$ and $\angle CDB = \angle CEA$ (corresponding angles between parallel lines)

Also $\angle C$ is common to triangles CBD and CAE.
Since the angles in triangle CBD are the same as in triangle CAE the triangles are similar.

Hence, by proportion: $\dfrac{CB}{CA} = \dfrac{CD}{CE} \left(= \dfrac{BD}{AE}\right)$

i.e. $\qquad \dfrac{9}{6+9} = \dfrac{CD}{12}$, from which $CD = 12\left(\dfrac{9}{15}\right) = \textbf{7.2 cm}$

Also, $\qquad \dfrac{9}{15} = \dfrac{BD}{10}$, from which $BD = 10\left(\dfrac{9}{15}\right) = \textbf{6 cm}$

*Problem 25* A rectangular shed 2 m wide and 3 m high stands against a perpendicular building of height 5.5 m. A ladder is used to gain access to the roof of the building. Determine the minimum distance between the bottom of the ladder and the shed and also the minimum length of ladder required.

A side view is shown in *Fig 26*, where AF is the minimum length of ladder. Since BD and CF are parallel, $\angle ADB = \angle DFE$ (corresponding angles between parallel lines). Hence triangles BAD and EDF are similar since their angles are the same.

AB = AC−BC = AC−DE = 5.5−3 = 2.5 m

By proportion: $\dfrac{AB}{DE} = \dfrac{BD}{EF}$, i.e. $\dfrac{2.5}{3} = \dfrac{2}{EF}$

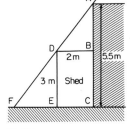

Hence **EF** = $2\left(\dfrac{3}{2.5}\right)$ = **2.4 m**

= **minimum distance from bottom of**
  **ladder to the shed.**

Since AC = 5.5 m, CF = BD+EF = 2+2.4 = 4.4 m,
then AF may be found using Pythagoras' theorem:
$AF^2 = 5.5^2 + 4.4^2$
**Hence minimum length of ladder,**
**AF** = $\sqrt{(5.5^2 + 4.4^2)}$ = **7.04 m**

**Fig 26**

*Further problems on similar triangles may be found in section C, problems 26 to 30, page 130.*

CONSTRUCTION OF TRIANGLES

*Problem 26* Construct a triangle whose sides are 6 cm, 5 cm and 3 cm

With reference to *Fig 27*:

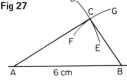

**Fig 27**

  (i) Draw a straight line of any length, and with a pair of compasses, mark out 6 cm length and label it AB.
 (ii) Set compass to 5 cm and with centre at A describe arc DE.
(iii) Set compass to 3 cm and with centre at B describe arc FG.
 (iv) The intersection of the two curves at C is the vertex of the required triangle. Join AC and BC by straight lines.

It may be proved by measurement that the ratio of the angles of a triangle is not equal to the ratio of the sides (i.e. in this problem, the angle opposite the 3 cm side **is not** equal to half the angle opposite the 6 cm side).

*Problem 27* Construct a triangle ABC such that $a$ = 6 cm, $b$ = 3 cm and $\angle C$ = 60°

With reference to *Fig 28*.

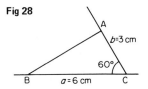

**Fig 28**

  (i) Draw a line BC, 6 cm long.
 (ii) Using a protractor centred at C make an angle of 60° to BC.
(iii) From C measure a length of 3 cm and label A.
 (iv) Join B to A by a straight line.

*Problem 28* Construct a triangle PQR given that QR = 5 cm, $\angle Q = 70°$ and $\angle R = 44°$

With reference to *Fig 29*:
(i) Draw a straight line 5 cm long and label it QR.
(ii) Use a protractor centred at Q and make an angle of 70°. Draw QQ'.
(iii) Use a protractor centred at R and make an angle of 44°. Draw RR'.
(iv) The intersection of QQ' and RR' forms the vertex P of the triangle. Join QP and RP by straight lines.

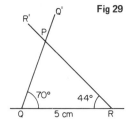

Fig 29

*Problem 29* Construct a triangle XYZ given that XY = 5 cm, the hypotenuse YZ = 6.5 cm and $\angle X = 90°$

With reference to *Fig 30*:
(i) Draw a straight line 5 cm long and label it XY.
(ii) Produce XY any distance to B. With compass centred at X make an arc at A and A'. (The length XA and XA' is arbitrary.) With compass centred at A draw the arc PQ. With the same compass setting and centred at A', draw the arc RS. Join the intersection of the arcs, C, to X, and a right angle to XY is produced at X. (Alternatively, a protractor can be used to construct a 90° angle.)

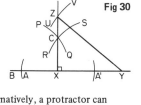

Fig 30

(iii) The hypotenuse is always opposite the right angle. Thus YZ is opposite $\angle X$. Using a compass centred at Y and set to 6.5 cm, describe the arc UV.
(iv) The intersection of the arc UV with XC produced, forms the vertex Z of the required triangle. Join YZ by a straight line.

*Further problems on constructions of triangles may be found in section C, Problem 31, page 131.*

(h) CIRCLES

*Problem 30* Find the circumference of a circle of radius 12.0 cm

Circumference, $c = 2 \times \pi \times \text{radius} = 2\pi r = 2\pi(12.0) = $ **75.40 cm**

*Problem 31* If the diameter of a circle is 75 mm, find its circumference

Circumference, $c = \pi \times \text{diameter} = \pi d = \pi(75) = $ **235.6 mm**

*Problem 32* Determine the radius of a circle if its perimeter is 112 cm

Perimeter = circumference, $c = 2\pi r$.

Hence $r = \dfrac{c}{2\pi} = \dfrac{112}{2\pi} = $ **17.83 cm**

*Problem 33* In *Fig 31*, AB is a tangent to the circle at B. If the circle radius is 40 mm and AB = 150 mm, calculate the length AO.

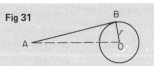

**Fig 31**

A tangent to a circle is at right angles to a radius drawn from the point of contact, i.e. $\angle ABO = 90°$

Hence, using Pythagoras' theorem:
$$AO^2 = AB^2 + OB^2$$
$$AO = \sqrt{(AB^2 + OB^2)} = \sqrt{[(150)^2 + (40)^2]}$$
$$= \textbf{155.2 mm}$$

*Problem 34* Convert to radians: (a) 125°; (b) 69° 47'

(a) Since $180° = \pi$ rads then $1° = \dfrac{\pi}{180}$ rads

Therefore $125° = 125\left(\dfrac{\pi}{180}\right)^c = $ **2.182 radians**

(Note that 'c' means 'circular measure' and indicates radian measure)

(b) $69° \ 47' = 69\dfrac{47}{60}° = 69.783°$

$69.783° = 69.783\left(\dfrac{\pi}{180}\right)^c = $ **1.218 radians**

*Problem 35* Convert to degrees and minutes: (a) 0.749 radians; (b) $\dfrac{3\pi}{4}$ radians

(a) Since $\pi$ rads = 180° then 1 rad = $\dfrac{180°}{\pi}$

Therefore $0.749$ rads = $0.749\left(\dfrac{180}{\pi}\right)° = 42.915°$

$0.915° = (0.915 \times 60)' = 55'$, correct to the nearest minute

**Hence 0.749 radians = 42° 55'**

(b) Since 1 rad = $\left(\dfrac{180}{\pi}\right)°$ then $\dfrac{3\pi}{4}\left(\dfrac{180}{\pi}\right)° = \dfrac{3}{4}(180)° = $ **135°**

*Problem 36* Express in radians, in terms of $\pi$:
(a) 45°; (b) 60°; (c) 90°; (d) 150°; (e) 270°; (f) 37.5°.

Since $180° = \pi$ rads then $1° = \dfrac{\pi}{180}$ rads

Hence:

(a) $45° = 45 \left(\dfrac{\pi}{180}\right)$ rads $= \dfrac{\pi}{4}$ **rads**

(b) $60° = 60 \left(\dfrac{\pi}{180}\right)$ rads $= \dfrac{\pi}{3}$ **rads**

(c) $90° = 90 \left(\dfrac{\pi}{180}\right)$ rads $= \dfrac{\pi}{2}$ **rads**

(d) $150° = 150 \left(\dfrac{\pi}{180}\right)$ rads $= \dfrac{5\pi}{6}$ **rads**

(e) $270° = 270 \left(\dfrac{\pi}{180}\right)$ rads $= \dfrac{3\pi}{2}$ **rads**

(f) $37.5° = 37.5 \left(\dfrac{\pi}{180}\right)$ rads $= \dfrac{75\pi}{360}$ rads $= \dfrac{5\pi}{24}$ **rads**

---

*Problem 37* Find the length of arc of a circle of radius 5.5 cm when the angle subtended at the centre is 1.20 radians.

Length of arc, $l = r\theta$, where $\theta$ is in radians
Hence $\qquad l = (5.5)(1.20) = $ **6.60 cm**

---

*Problem 38* Determine the diameter and circumference of a circle if an arc of length 4.75 cm subtends an angle of 0.91 radians

Since $l = r\theta$ then $r = \dfrac{l}{\theta} = \dfrac{4.75}{0.91} = 5.22$ cm

Diameter $= 2 \times$ radians $= 2 \times 5.22 = $ **10.44 cm**
Circumference, $c = \pi d = \pi(10.44) = $ **32.80 cm**

---

*Problem 39* If an angle of 125° is subtended by an arc of a circle of radius 8.4 cm, find the length of (a) the minor arc, and (b) the major arc, correct to 3 significant figures.

Since $180° = \pi$ rads, then $1° = \left(\dfrac{\pi}{180}\right)$ rads and $125° = 125 \left(\dfrac{\pi}{180}\right)$ rads

Length of minor arc, $l = r\theta = (8.4)(125)\left(\dfrac{\pi}{180}\right) = $ **18.3 cm, correct to 3 significant figures**

126

Length of major arc = (circumference−minor arc) = $2\pi(8.4)−18.3 =$ **34.5 cm,**
**correct to 3 significant figures**
(Alternatively, major arc = $r\theta = 8.4(360−125) \left(\dfrac{\pi}{180}\right) = 34.5$ cm)

---

*Problem 40* Determine the angle, in degrees and minutes, subtended at the centre of a circle of diameter 42 mm by an arc of length 36 mm.

Since length of arc, $l = r\theta$ then $\theta = \dfrac{l}{r}$

Radius, $r = \dfrac{\text{diameter}}{2} = \dfrac{42}{2} = 21$ mm

Hence $\theta = \dfrac{l}{r} = \dfrac{36}{21} = 1.7143$ radians

$1.7143$ rads $= 1.7143 \times \left(\dfrac{180}{\pi}\right)^{\circ} = 98.22° = $ **98° 13′ = angle subtended at centre**
**of circle**

---

*Problem 41* If an arc of length 11.48 cm subtends an angle of 168° 27′ at the centre of a circle, find its radius correct to the nearest millimetre.

$168° 27′ = 168\dfrac{27}{60}^{\circ} = 168.45° = 168.45 \left(\dfrac{\pi}{180}\right)$ radians

Hence $\theta = 2.94$ radians

Since arc length $l = r\theta$ then $r = \dfrac{l}{\theta} = \dfrac{11.48}{2.94} = 3.905$ cm

$= 39.05$ mm

$=$ **39 mm to the nearest millimetre**

*Further problems on circles may be found in section C following, problems 32 to 46, page 131.*

---

## C. FURTHER PROBLEMS ON GEOMETRY

*Angular measurement*

1  Add together the following angles:
    (a) 32° 19′ and 49° 52′;               (b) 29° 42′, 56° 37′ and 63° 54′;
    (c) 21° 33′ 27″ and 78° 42′ 36″;        (d) 48° 11′ 19″, 31° 41′ 27″ and 9° 9′ 37″.
              [(a) 82° 11′; (b) 150° 13′; (c) 100° 16′ 3″; (d) 89° 2′ 23″]
2  Determine (a) 17°−9° 49′;                (b) 43° 37′− 15° 49′;
    (c) 78° 29′ 41″−59° 41′ 52″;            (d) 114°−47° 52′ 37″
              [(a) 7° 11′; (b) 27° 48′; (c) 18° 47′ 49″; (d) 66° 7′ 23″]
3  Convert the following angles to degrees and decimals of a degree, correct to
    3 decimal places:
    (a) 15° 11′; (b) 29° 53′; (c) 49° 42′ 17″; (d) 135° 7′ 19″
              [(a) 15.183°; (b) 29.883°; (c) 49.705°; (d) 135.122°]

4   Convert the following angles into degrees, minutes and seconds:
    (a) 25.4°; (b) 36.48°; (c) 55.724°; (d) 231.025°
            [(a) 25° 24′ 0″; (b) 36° 28′ 48″; (c) 55° 43′ 26″; (d) 231° 1′ 30″]

*Types and properties of angles*

5   State the general name given to the following angles:
    (a) 63°; (b) 147°; (c) 250°

            [(a) acute; (b) obtuse; (c) reflex]

6   Determine the angles complementary to the following:
    (a) 69°; (b) 27° 37′; (c) 41° 3′ 43″

            [(a) 21°; (b) 62° 23′; (c) 48° 56′ 17″]

7   Determine the angles supplementary to the following:
    (a) 78°; (b) 15°; (c) 169° 41′ 11″

            [(a) 102°; (b) 165°; (c) 10° 18′ 49″]

8   With reference to *Fig 32*, what is the name given to the line XY.
    Give examples of each of the following:
    (a) vertically opposite angles:        (b) supplementary angles;
    (c) corresponding angles;              (d) alternate angles.

    ⎡ Transversal; (a) 1 & 3, 2 & 4, 5 & 7, 6 & 8; ⎤
    ⎢ (b) 1 & 2, 2 & 3, 3 & 4, 4 & 1, 5 & 6,      ⎥
    ⎢      6 & 7, 7 & 8, 8 & 5, 3 & 8, 1 & 6,      ⎥
    ⎢      4 & 7 or 2 & 5;                         ⎥
    ⎢ (c) 1 & 5, 2 & 6, 4 & 8, 3 & 7;             ⎥
    ⎣ (d) 3 & 5 or 2 & 8                          ⎦

9   In *Fig 33*, find angle $\alpha$.

            [59° 20′]

10  In *Fig 34*, find angles *a*, *b* and *c*

            [$a = 69°$, $b = 21°$, $c = 82°$]

11  Find angle $\beta$ in *Fig 35*.

            [51°]

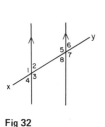

**Fig 32**

**Fig 34**

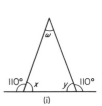

(i)

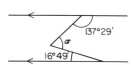

**Fig 33**        **Fig 35**

**Fig 36**

128

*Properties of triangles*

12 In *Fig 36, (i) and (ii)*, find angles *w, x, y* and *z*. What is the name given to the types of triangle shown in both (*i*) and (*ii*)?

[40°, 70°, 70°, 125°; isosceles]

13 Find the values of angles *a* to *g* in *Fig 37(i) and (ii)*.

$$\begin{bmatrix} a = 18° \ 50', b = 71° \ 10'; \\ c = 68°, d = 90°, e = 22°; \\ f = 49°, g = 41° \end{bmatrix}$$

14 Find the unknown angles *a* to *k* in *Fig 38*.

$$\begin{bmatrix} a = 103°, b = 55°, c = 77°; \\ d = 125°, e = 55°, f = 22°; \\ g = 103°, h = 77°, i = 103°; \\ j = 77°, k = 81° \end{bmatrix}$$

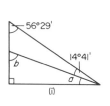

(i)

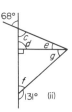

**Fig 37**

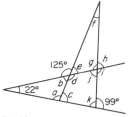

**Fig 38**

**Fig 39**

15 Triangle ABC has a right angle at B and ∠BAC is 34°. BC is produced to D. If the bisectors of ∠ABC and ∠ACD meet at E, determine ∠BEC.

[17°]

16 If in *Fig 39*, triangle BCD is equilateral, find the interior angles of triangle ABE.

[*A* = 37°, *B* = 60°, *E* = 83°]

*Theorem of Pythagoras*

17 In a triangle ABC, ∠B is a right angle, AB = 6.92 cm and BC = 8.78 cm. Find the length of the hypotenuse.

[11.18 cm]

18 In a triangle CDE, ∠D = 90°, CD = 14.83 mm and CE = 28.31 mm. Determine the length of DE.

[24.11 mm]

19 Show that if a triangle has sides of 8, 15 and 17 cm it is right-angled.

20 Triangle PQR is isosceles, ∠Q being a right angle. If the hypotenuse is 38.47 cm find (a) the lengths of sides PQ and QR, and (b) the value of ∠QPR.

[(a) 27.20 cm each; (b) 45°]

129

21 A man cycles 24 km due south and then 20 km due east. Another man, starting at the same time as the first man, cycles 32 km due east and then 7 km due south. Find the distance between the two men. [20.81 km]

22 A ladder 3.5 m long is placed against a perpendicular wall with its foot 2.0 m from the wall. How far up the wall (to the nearest cm) does the ladder reach? If the foot of the ladder is now moved 50 cm further away from the wall, how far does the top of the ladder fall?

[2.87 m; 42 cm]

23 Two ships leave a port at the same time. One travels due west at 18.4 km/h and the other due south at 27.6 km/h. Calculate how far apart the two ships are after 4 hours.

[132.7 km]

*Congruent triangles*

24 State which of the pairs of triangles in *Fig 40* are congruent and name their sequence.

$$\begin{bmatrix} \text{(a) Congruent BAC, DAC (SAS)} \\ \text{(b) Congruent FGE, JHI (SSS)} \\ \text{(c) Not necessarily congruent} \\ \text{(d) Congruent QRT, SRT (RHS)} \\ \text{(e) Congruent UVW, XZY (ASA)} \end{bmatrix}$$

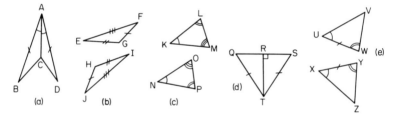

**Fig 40**

25 In a triangle ABC, AB = BC and D and E are points on AB and BC respectively such that AD = CE. Show that triangles AEB and CDB are congruent.

*Similar triangles*

26 In *Fig 41*, find the lengths $x$ and $y$.

[$x = 16.54$ mm; $y = 4.18$ mm]

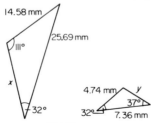

**Fig 41**

27 PQR is an equilateral triangle of side 4 cm. When PQ and PR are produced to S and T respectively ST is found to be parallel with QR. If PS is 9 cm, find the length of ST. X is a point on ST between S and T such that the line PX is the bisector of ∠SPT. Find the length of PX.

[9 cm; 7.79 cm]

28 The triangle ABC is right angled at B. If AB = 4 cm and BC = 3 cm find the length of BD, which is the length of the perpendicular to the hypotenuse, and also the length AD and DC.

[BD = 2.4 cm; AD = 3.2 cm; DC = 1.8 cm]

**Fig 42**

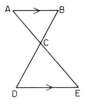

29 In *Fig 42*, find (a) the length of BC when AB = 6 cm, DE = 8 cm and DC = 3 cm, (b) the length of DE when EC = 2 cm, AC = 5 cm and AB = 10 cm.

[(a) 2.25 cm; (b) 4 cm]

30 In *Fig 43*, AF = 8 m, AB = 5 m and BC = 3 m. Find the length of BD. [3 m]

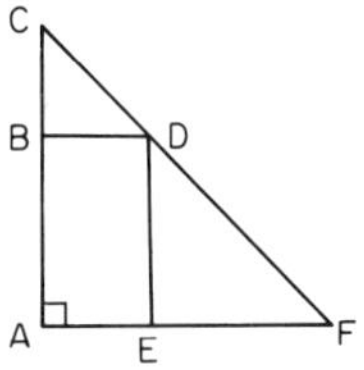

**Fig 43**

*Construction of triangles*

31 Construct triangles ABC, given:
   (a) *a* = 8 cm, *b* = 6 cm and *c* = 5 cm; (b) *a* = 40 mm, *b* = 60 mm and *C* = 60°;
   (c) *a* = 6 cm, *C* = 45° and *B* = 75°; (d) *c* = 4 cm, *A* = 130° and *C* = 15°;
   (e) *a* = 90 mm, *B* = 90°, hypotenuse = 105 mm.

*Circles*

32 Calculate the length of the circumference of a circle of radius 7.2 cm.

[45.24 cm]

33 If the diameter of a circle is 82.6 mm, calculate the circumference of the circle.

[259.5 mm]

34 Determine the radius of a circle whose circumference is 16.52 cm.

[2.629 cm]

35 Find the diameter of a circle whose perimeter is 149.8 cm.

[47.68 cm]

36 Convert to radians in terms of π; (a) 30°; (b) 75°; (c) 225°

$$\left[ \text{(a)} \frac{\pi}{6}; \text{(b)} \frac{5\pi}{12}; \text{(c)} \frac{5\pi}{4} \right]$$

37 Convert to radians: (a) 48°; (b) 84° 51′; (c) 232° 15′.

[(a) 0.838; (b) 1.481; (c) 4.054]

38 Convert to degrees: (a) $\frac{5\pi}{6}$ rads, (b) $\frac{4\pi}{9}$ rads, (c) $\frac{7\pi}{12}$ rads.

[(a) 150°; (b) 80°; (c) 105°]

39 Convert to degrees and minutes: (a) 0.0125 rads, (b) 2.69 rads, (c) 7.241 rads.

[(a) 0° 43′; (b) 154° 8′; (c) 414° 53′]

40 Find the length of an arc of a circle of radius 8.32 cm when the angle subtended at the centre is 2.14 radians.

[17.80 cm]

41 If the angle subtended at the centre of a circle of diameter 82 mm is 1.46 rads, find the lengths of the (a) minor arc, (b) major arc.

[(a) 59.86 mm; (b) 197.8 mm]

42 A pendulum of length 1.5 m swings through an angle of 10° in a single swing. Find, in centimetres, the length of the arc traced by the pendulum bob.

[26.2 cm]

43 Determine the length of the radius and circumference of a circle if an arc length of 32.6 cm subtends an angle of 3.76 radians.

[8.67 cm; 54.48 cm]

44 An arc subtends an angle of 96° at the centre of a circle of radians 125 mm. Find the length of the arc.

[209.4 mm]

45 Determine the angle of lap, in degrees and minutes, if 180 mm of a belt drive are in contact with a pulley of diameter 250 mm.

[82° 30 ]

46 Determine the number of complete revolutions a motorcycle wheel will make in travelling 2 km, if the wheel's diameter is 85.1 cm.

[748]

# 10 Areas and volumes

## A. MAIN POINTS CONCERNED WITH AREAS AND VOLUMES

1 **Mensuration** is a branch of mathematics concerned with the determination of lengths, areas and volumes.

2 A **polygon** is a closed plane figure bounded by straight lines. A polygon which has:
   (i) 3 sides is called a **triangle**
   (ii) 4 sides is called a **quadrilateral**
   (iii) 5 sides is called a **pentagon**
   (iv) 6 sides is called a **hexagon**
   (v) 7 sides is called a **heptagon**
   (vi) 8 sides is called a **octagon**.

3 There are five types of **quadrilateral**, these being: (i) rectangle, (ii) square, (iii) parallelogram, (iv) rhombus, (v) trapezium. (The properties of these are given in paragraphs 4 to 8.) If the opposite corners of any quadrilateral are joined by a straight line, two triangles are produced. Since the sum of the angles of a triangle is 180°, the sum of the angles of a quadrilateral is 360°.

4 In a **rectangle**, shown in *Fig 1*:
   (i) all four angles are right angles,
   (ii) opposite sides are parallel and equal in length, and
   (iii) diagonals AC and BD are equal in length and bisece one another.

Fig 1

5 In a **square**, shown in *Fig 2*:
   (i) all four angles are right angles,
   (ii) opposite sides are parallel,
   (iii) all four sides are equal in length, and
   (iv) diagonals PR and QS are equal in length and bisect one another at right angles.

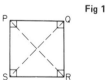

Fig 2

6 In a **parallelogram**, shown in *Fig 3*:
   (i) opposite angles are equal,
   (ii) opposite sides are parallel and equal in length, and
   (iii) diagonals WY and XZ bisect one another.

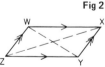

Fig 3

7 In a **rhombus**, shown in *Fig 4*:
   (i) opposite angles are equal,
   (ii) opposite angles are bisected by a diagonal,
   (iii) opposite sides are parallel,
   (iv) all four sides are equal in length, and
   (v) diagonals AC and BD bisect one another at right angles.

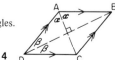

Fig 4

o In a **trapezium**, shown in *Fig 5*:
  (i) only one pair of sides is parallel.

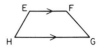

**Fig 5**

## 9 Areas of plane figures

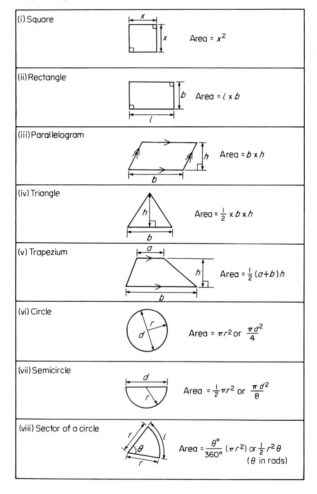

## 10 Volumes and surface area of regular solids

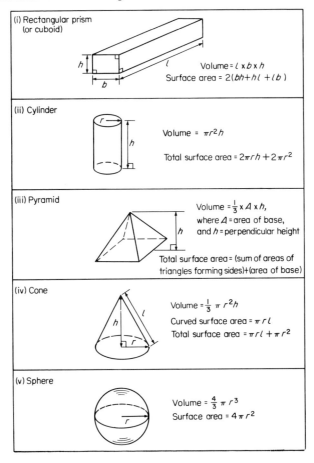

**(i) Rectangular prism (or cuboid)**

Volume = $l \times b \times h$
Surface area = $2(bh + hl + lb)$

**(ii) Cylinder**

Volume = $\pi r^2 h$

Total surface area = $2\pi rh + 2\pi r^2$

**(iii) Pyramid**

Volume = $\frac{1}{3} \times A \times h$,
where $A$ = area of base,
and $h$ = perpendicular height

Total surface area = (sum of areas of triangles forming sides) + (area of base)

**(iv) Cone**

Volume = $\frac{1}{3} \pi r^2 h$
Curved surface area = $\pi r l$
Total surface area = $\pi r l + \pi r^2$

**(v) Sphere**

Volume = $\frac{4}{3} \pi r^3$
Surface area = $4\pi r^2$

11 (i) **The areas of similar shapes are proportional to the squares of corresponding linear dimensions.**
For example, *Fig 6* shows two squares, one of which has sides three times as long as the other.
Area of fig. 6(a) = $(x)(x)$ = $x^2$
Area of fig. 6(b) = $(3x)(3x)$ = $9x^2$
Hence fig. 6(b) has an area $(3)^2$, i.e. 9 times the area of fig. 6(a).
(See *Problem 7.*)

**Fig 6**

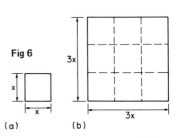

(a)          (b)

(ii) **The volumes of similar bodies are proportional to the cubes of corresponding linear dimensions.**
For example, *Fig 7* shows two cubes, one of which has sides three times as long as those of the other.
Volume of fig. 7(a) = $(x)(x)(x)$ = $x^3$
Volume of fig. 7(b) = $(3x)(3x)(3x) = 27x^3$
Hence *Fig 7(b)* has a volume $(3)^3$, i.e. 27 times the volume of *Fig 7(a)*.
(See *Problem 17*.)

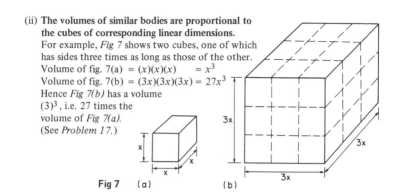

**Fig 7**   (a)   (b)

## B. WORKED PROBLEMS ON AREAS AND VOLUMES

### (a) AREAS OF PLANE FIGURES

*Problem 1* State the types of quadrilateral shown in *Fig 8* and determine the angles marked *a* to *l*.

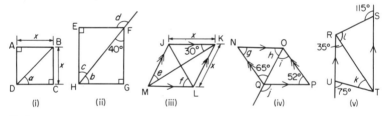

**Fig 8**

(i) *ABCD is a square*
The diagonals of a square bisect each of the right angles
Hence $a = \dfrac{90°}{2} = \mathbf{45°}$

(ii) *EFGH is a rectangle*
In triangle FGH, $40°+90°+b = 180°$ (angles in a triangle add up to 180°)
from which, $b = \mathbf{50°}$
Also $c = \mathbf{40°}$ (alternate angles between parallel lines EF and HG)
(Alternatively, *b* and *c* are complementary, i.e. add up to 90°)
$d = 90°+c$ (external angle of a triangle equals the sum of the interior opposite angles)
Hence $d = 90°+40° = \mathbf{130°}$

(iii) *JKLM is a rhombus*

The diagonals of a rhombus bisect the interior angles and opposite internal angles are equal.

Thus $\angle$JKM = $\angle$MKL = $\angle$JMK = $\angle$LMK = 30°

Hence $e$ = **30°**

In triangle KLM, 30°+$\angle$KLM+30° = 180° (angles in a triangle add up to 180°)

Hence $\angle$KLM = 120°

The diagonal JK bisects $\angle$KLM, hence $f = \dfrac{120°}{2}$ = **60°**

(iv) *NOPQ is a parallelogram*

$g$ = **52°** (since opposite interior angles of a parallelogram are equal). In triangle NOQ, $g+h+65°$ = 180° (angles in a triangle add up to 180°) from which,

$h = 180°-65°-52°$ = **63°**.

$i$ = **65°** (alternate angles between parallel lines NQ and OP)

$j = 52°+i = 52°+65°$ = **117°** (external angle of a triangle equals the sum of the interior opposite angles)

(v) *RSTU is a trapezium*

$35°+k = 75°$ (external angle of a triangle equals the sum of the interior opposite angles)

Hence $k$ = **40°**

$\angle$STR = 35° (alternate angles between parallel lines RU and ST)

$l+35° = 115°$ (external angle of a triangle equals the sum of the interior opposite angles)

Hence $l = 115°-35°$ = **80°**

---

*Problem 2* A rectangular tray is 820 mm long and 400 mm wide. Find its area in (a) mm², (b) cm², (c) m².

(a) Area = length × width = 820 × 400 = **328 000 mm²**

(b) 1 cm² = 100 mm². Hence 328 000 mm² = $\dfrac{328\ 000\ \text{cm}^2}{100}$ = **3280 cm²**

(c) 1 m² = 10 000 cm². Hence 3280 cm² = $\dfrac{3280}{10\ 000}$ m² = **0.3280 m²**

*Problem 3* Find (a) the cross-sectional area of the girder shown in *Fig 9(a)*, and (b) the area of the path shown in *Fig 9(b)*.

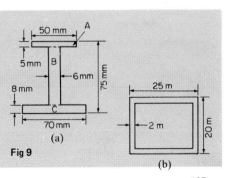

(a)

(b)

**Fig 9**

(a) The girder may be divided into three separate rectangles as shown.
Area of rectangle A = 50 × 5 = 250 mm²
Area of rectangle B = (75−8−5) × 6 = 62 × 6 = 372 mm²
Area of rectangle C = 70 × 8 = 560 mm²
Total area of girder = 250+372+560 = **1182 mm²** or **11.82 cm²**

(b) Area of path = area of large rectangle−area of small rectangle
= (25 × 20)−(21 × 16)
= 500−336 = **164 m²**

---

*Problem 4* Find the area of the parallelogram shown in *Fig 10*, (dimensions are in mm)

---

Area of parallelogram = base × perpendicular height.
The perpendicular height $h$ is found using Pythagoras' theorem.

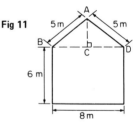

**Fig 10**

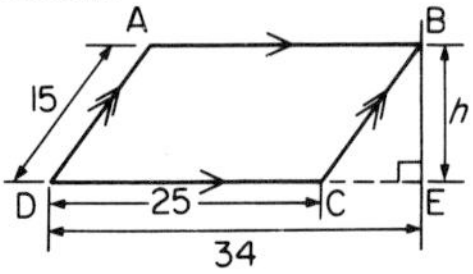

$BC^2 = CE^2+h^2$
i.e. $15^2 = (34−25)^2+h^2$
$h^2 = 15^2−9^2 = 225−81 = 144$
Hence, $h = \sqrt{144} = 12$ mm ($−12$ can be neglected)
Hence, area of ABCD = 25 × 12 = **300 mm²**

---

*Problem 5* *Fig 11* shows the gable end of a building. Determine the area of brickwork in the gable end.

---

The shape is that of a rectangle and a triangle.
Area of rectangle = 6 × 8 = 48 m²

Area of triangle = $\frac{1}{2}$× base × height

CD = 4 m, AD = 5 m. Hence AC = 3 m
(since it is a 3,4,5 triangle)

Hence, area of triangle ABD = $\frac{1}{2}$× 8 × 3 = 12 m²
Total area of brickwork = 48+12 = **60 m²**

**Fig 11**

---

*Problem 6* Determine the area of the shape shown in *Fig 12*

---

The shape shown is a trapezium.

Area of trapezium = $\frac{1}{2}$(sum of parallel sides)(perpendicular distance between them)

**Fig 12**

$= \frac{1}{2}(27.4+8.6)(5.5)$

$= \frac{1}{2}$× 36 × 5.5 = **99 mm²**

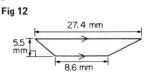

*Problem 7* A rectangular garage is shown on a building plan having dimensions 10 mm by 20 mm. If the plan is drawn to a scale of 1 to 250, determine the true area of the garage in square metres.

Area of garage on the plan = 10 mm × 20 mm = 200 mm$^2$.
Since the areas of similar shapes are proportional to the squares of corresponding dimensions then:
True area of garage = 200 × (250)$^2$ = 12.5 × 10$^6$ mm$^2$
$$= \frac{12.5 \times 10^6}{10^6} \text{ m}^2 = \textbf{12.5 m}^2$$

*Problem 8* Find the areas of the circles having (a) a radius of 5 cm, (b) a diameter of 15 mm, (c) a circumference of 70 mm

Area of a circle $= \pi r^2$ or $\frac{\pi d^2}{4}$

(a) Area $= \pi r^2 = \pi(5)^2 = 25\pi = \textbf{78.54 cm}^2$

(b) Area $= \frac{\pi d^2}{4} = \frac{\pi(15)^2}{4} = \frac{225\pi}{4} = \textbf{176.7 mm}^2$

(c) Circumference, $c = 2\pi r$ (from chapter 9)

Hence $r = \frac{c}{2\pi} = \frac{70}{2\pi} = \frac{35}{\pi}$ mm

Area of circle $= \pi r^2 = \pi \left(\frac{35}{\pi}\right)^2 = \frac{35^2}{\pi} = \textbf{389.9 mm}^2$ or $\textbf{3.899 cm}^2$

*Problem 9* Calculate the areas of the following sectors of circles:
(a) having radius 6 cm with angle subtended at centre 50°,
(b) having diameter 80 mm with angle subtended at centre 107° 42′,
(c) having radius 8 cm with angle subtended at centre 1.15 radians.

Area of sector of a circle $= \frac{\theta°}{360}(\pi r^2)$ or $\frac{1}{2}r^2\theta$ ($\theta$ in radians)

(a) Area of sector $= \frac{50}{360}(\pi 6^2) = \frac{50 \times \pi \times 36}{360} = 5\pi = \textbf{15.71 cm}^2$

(b) If diameter = 80 mm, then radius, $r = 40$ mm.

Area of sector $= \frac{107° 42′}{360}(\pi 40^2) = \frac{107\frac{42}{60}}{360}(\pi 40^2) = \frac{107.7}{360}(\pi 40^2)$
$$= \textbf{1504 mm}^2 \text{ or } \textbf{15.04 cm}^2$$

(c) Area of sector $= \frac{1}{2}r^2\theta = \frac{1}{2} \times 8^2 \times 1.15 = \textbf{36.8 cm}^2$

*Problem 10* A hollow shaft has an outside diameter of 5.45 cm and an inside diameter of 2.25 cm. Calculate the cross-sectional area of the shaft.

Fig 13

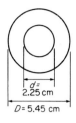

The cross-sectional area of the shaft is shown by the shaded part in *Fig 13* (often called an **annulus**).

Area of shaded part = area of large circle−area of small circle

$$= \frac{\pi D^2}{4} - \frac{\pi d^2}{4} = \frac{\pi}{4}(D^2 - d^2) = \frac{\pi}{4}(5.45^2 - 2.25^2) = 19.35 \text{ cm}^2$$

$d =$ 2.25 cm

$D =$ 5.45 cm

*Problem 11* Calculate the area of a regular octagon, if each side is 5 cm and the width across the flats is 12 cm.

An octagon is a 8-sided polygon. If radii are drawn from the centre of the polygon to the vertices then 8 equal triangles are produced (see *Fig 14*).

Area of one triangle $= \frac{1}{2} \times$ base $\times$ height

$$= \frac{1}{2} \times 5 \times \frac{12}{2} = 15 \text{ cm}^2$$

Area of octagon $= 8 \times 15 = \textbf{120 cm}^2$

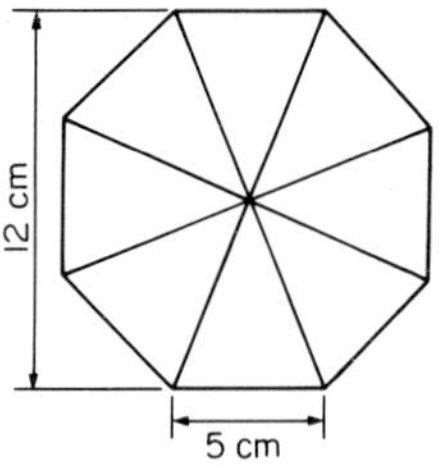

12 cm

5 cm

**Fig 14**

*Problem 12* Determine the area of a regular hexagon which has sides 8 cm long.

A hexagon is a 6-sided polygon which may be divided into 6 equal triangles as shown in *Fig 15*. The angle subtended at the centre of each triangle is $\frac{360}{6} = 60°$. The other two angles in the triangle add up to 120° and are equal to each other. Hence each of the triangles is equilateral with each angle 60° and each side 8 cm.

Area of one triangle $= \frac{1}{2} \times$ base $\times$ height $= \frac{1}{2} \times 8 \times h$.

$h$ is calculated using Pythagoras' theorem: $8^2 = h^2 + 4^2$

from which $h = \sqrt{(8^2 - 4^2)}$

$$= 6.928 \text{ cm}$$

Hence area of one triangle $= \frac{1}{2} \times 8 \times 6.928 = 27.71 \text{ cm}^2$

Area of hexagon $= 6 \times 27.71 = \textbf{166.3 cm}^2$

**Fig 15**

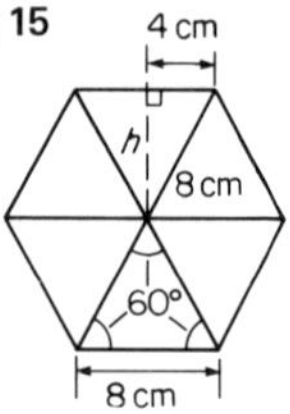

4 cm

$h$

8 cm

60°

8 cm

*Further problems on areas of plane figures may be found in section C, problems 1 to 15, page 146.*

**Problem 13** A water tank is the shape of a rectangular prism having length 2 m, breadth 75 cm and height 50 cm. Determine the capacity of the tank in (a) m³; (b) cm³; (c) litres.

Volume of rectangular prism = $l \times b \times h$ (see table on page 135).
(a) Volume of tank = $2 \times 0.75 \times 0.5 =$ **0.75 m³**
(b) $1$ m³ $= 10^6$ cm³. Hence $0.75$ m³ $= 0.75 \times 10^6$ cm³ $=$ **750 000 cm³**
(c) $1$ litre $= 1000$ cm³. Hence $750\ 000$ cm³ $= \dfrac{750\ 000}{1000}$ litres $=$ **750 litres**

**Problem 14** Find the volume and total surface area of a cylinder of length 15 cm and diameter 8 cm.

Volume of cylinder = $\pi r^2 h$ (see table on page 135).
Since diameter = 8 cm, then radius, $r = 4$ cm
Hence volume = $\pi \times 4^2 \times 15 =$ **754 cm³**
Total surface area (i.e. including the two ends) = $2\pi rh + 2\pi r^2$
$\qquad\qquad\qquad\qquad\qquad\qquad\qquad\qquad = (2 \times \pi \times 4 \times 15) + (2 \times \pi \times 4^2)$
$\qquad\qquad\qquad\qquad\qquad\qquad\qquad\qquad =$ **477.5 cm²**

**Problem 15** Determine the volume (in cm³) of the shape shown in *Fig 16*.

The solid shown in *Fig 16* is a triangular prism.
The volume $V$ of any prism is given by: $V = Ah$, when $A$ is the cross-sectional area and $h$ is the perpendicular height.

Hence volume = $(\frac{1}{2} \times 16 \times 12) \times 40$
$\qquad\qquad = 3840$ mm³ $=$ **3.840 cm³**
$\qquad\qquad$ (since $1$ cm³ $= 1000$ mm³)

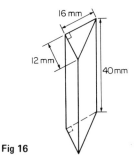

**Fig 16**

**Problem 16** Calculate the volume and total surface area of the solid prism shown in *Fig 17*.

The solid shown in *Fig 17* is a trapezoidal prism.
Volume = cross-sectional area $\times$ height
$\qquad = \left[\frac{1}{2}(11+5)4\right] \times 15 = 32 \times 15 =$ **480 cm³**

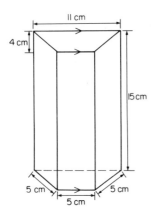

**Fig 17**

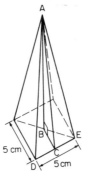

**Fig 18**

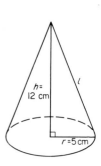

**Fig 19**

Surface area  = sum of two trapeziums + 4 rectangles
= $(2 \times 32)+(5 \times 15)+(11 \times 15)+2(5 \times 15)$
= $64+75+165+150 = $ **454 cm$^2$**

---

*Problem 17* A car has a mass of 1000 kg. A model of the car is made to a scale of 1 to 50. Determine the mass of the model if the car and its model are made of the same material.

$\dfrac{\text{Volume of model}}{\text{Volume of car}} = \left(\dfrac{1}{50}\right)^3$, since the volume of similar bodies are proportional

to the cube of corresponding dimensions.
Mass = density × volume, and since both car and model are made of the same material then:

$\dfrac{\text{Mass of model}}{\text{Mass of car}} = \left(\dfrac{1}{50}\right)^3$

Hence mass of model = (mass of car) $\left(\dfrac{1}{50}\right)^3 = \dfrac{1000}{50^3} = $ **0.008 kg or 8 g.**

---

*Problem 18* Determine the volume and the total surface area of the square pyramid shown in *Fig 18* if its perpendicular height is 12 cm.

Volume of pyramid = $\dfrac{1}{3}$(area of base) × perpendicular height

$= \dfrac{1}{3}(5 \times 5) \times 12 = $ **100 cm$^3$**

The total surface area consists of a square base and 4 equal triangles.

142

Area of triangle ADE = $\frac{1}{2} \times$ base $\times$ perpendicular height = $\frac{1}{2} \times 5 \times$ AC.

The length AC may be calculated using Pythagoras' theorem on triangle ABC, where AB = 12 cm, BC = $\frac{1}{2} \times 5$ = 2.5 cm.

$AC = \sqrt{(AB^2 + BC^2)} = \sqrt{(12^2 + 2.5^2)} = 12.26$ cm

Hence area of triangle ADE = $\frac{1}{2} \times 5 \times 12.26 = 30.65$ cm$^2$

Total surface area of pyramid = $(5 \times 5) + 4(30.65) = \textbf{147.6 cm}^2$

*Problem 19* Determine the volume and total surface area of a cone of radius 5 cm and perpendicular height 12 cm.

The cone is shown in *Fig 19*.

Volume of cone = $\frac{1}{3}\pi r^2 h = \frac{1}{3} \times \pi \times 5^2 \times 12 = \textbf{314.2 cm}^3$

Total surface area = curved surface area + area of base

$= \pi r l + \pi r^2$

From *Fig 19*, slant height $l$ may be calculated using Pythagoras' theorem

$l = \sqrt{(12^2 + 5^2)} = 13$ cm

Hence total surface area = $(\pi \times 5 \times 13) + (\pi \times 5^2) = \textbf{282.7 cm}^2$

*Problem 20* Find the volume and surface area of a sphere of diameter 8 cm.

Since diameter = 8 cm, then radius, $r = 4$ cm

Volume of sphere = $\frac{4}{3}\pi r^3 = \frac{4}{3} \times \pi \times 4^3 = \textbf{268.1 cm}^3$

Surface area of sphere = $4\pi r^2 = 4 \times \pi \times 4^2 = \textbf{201.1 cm}^2$

*Problem 21* A wooden section is shown in *Fig 20*. Find (a) its volume (in m$^3$); and (b) its total surface area.

The section of wood is a prism whose end comprises a rectangle and a semicircle. Since the radius of the semicircle is 8 cm, the diameter is 16 cm. Hence the rectangle has dimensions 12 cm by 16 cm.

Area of end = $(12 \times 16) + \frac{1}{2}\pi 8^2$

$= 292.5$ cm$^2$

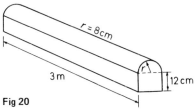

**Fig 20**

143

Volume of wooden section = area of end × perpendicular height

$$= 292.5 \times 300 = 87\ 750\ \text{cm}^3 = \frac{87\ 750\ \text{m}^3}{10^6}$$

$$= 0.087\ 75\ \text{m}^3$$

The total surface area comprises the two ends (each of area 292.5 cm$^2$), 3 rectangles and a curved surface (which is half a cylinder).

Hence total surface area = $(2 \times 292.5) + 2(12 \times 300) + (16 \times 300) + \frac{1}{2}(2\pi \times 8 \times 300)$

$$= 585 + 7200 + 4800 + 2400\pi$$

$$= \textbf{20 125 cm}^2 \text{ or } \textbf{2.0125 m}^2$$

*Problem 22* A rectangular piece of metal having dimensions 4 cm by 3 cm by 12 cm is melted down and recast into a pyramid having a rectangular base measuring 2.5 cm by 5 cm. Calculate the perpendicular height of the pyramid.

Volume of rectangular prism of metal = $4 \times 3 \times 12 = 144\ \text{cm}^3$.

Volume of pyramid = $\frac{1}{3}$(area of base)(perpendicular height)

Assuming no waste of metal, $144 = \frac{1}{3}(2.5 \times 5)$(height)

i.e. perpendicular height = $\dfrac{144 \times 3}{2.5 \times 5} = \textbf{34.56 cm}$

*Problem 23* A rivet consists of a cylindrical head, of diameter 1 cm and depth 2 mm, and a shaft of diameter 2 mm and length 1.5 cm. Determine the volume of metal in 2000 such rivets.

Radius of cylindrical head = $\frac{1}{2}$cm = 0.5 cm; height of cylindrical head = 2 mm

$$= 0.2\ \text{cm}$$

Hence, volume of cylindrical head = $\pi r^2 h = \pi(0.5)^2(0.2) = 0.1571\ \text{cm}^3$

Volume of cylindrical shaft = $\pi r^2 h = \pi\left(\dfrac{1}{10}\right)^2(1.5) = 0.0471\ \text{cm}^3$

Total volume of 1 rivet = 0.1571 + 0.0471 = 0.2042 cm$^3$
Volume of metal in 2000 such rivets = 2000 × 0.2042 = **408.4 cm$^3$**

*Problem 24* A solid metal cylinder of radius 6 cm and height 15 cm is melted down and recast into a shape comprising a hemisphere surmounted by a cone. Assuming that 8% of the metal is wasted in the process, determine the height of the conical portion, if its diameter is to be 12 cm.

Volume of cylinder = $\pi r^2 h = \pi \times 6^2 \times 15 = 540\pi\ \text{cm}^3$

If 8% of metal is lost then 92% of 540$\pi$ gives the volume of the new shape (shown in *Fig 21*).

144

Hence volume of (hemisphere+ cone) = $0.92 \times 540\pi$ cm$^3$      **Fig 21**

i.e. $\frac{1}{2}\left(\frac{4}{3}\pi r^3\right) + \frac{1}{3}\pi r^2 h = 0.92 \times 540\pi$

Dividing throughout by $\pi$ gives: $\frac{2}{3}r^3 + \frac{1}{3}r^2 h = 0.92 \times 540$

Since the diameter of the new shape is to be 12 cm, then radius, $r = 6$ cm.

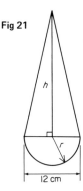

Hence $\frac{2}{3}(6)^3 + \frac{1}{3}(6)^2 h = 0.92 \times 540$

$144 + 12h = 496.8$

i.e. height of conical portion, $h = \dfrac{496.8 - 144}{12} = $ **29.4 cm**

12 cm

---

*Problem 25* A block of copper having a mass of 50 kg is drawn out to make 500 m of wire of uniform cross-section. Given that the density of copper is 8.91 g/cm$^3$, calculate (a) the volume of copper, (b) the cross-sectional area of the wire, and (c) the diameter of the cross-section of the wire.

(a) A density of 8.91 g/cm$^3$ means that 8.91 g of copper has a volume of 1 cm$^3$, or 1 g of copper has a volume of 1/8.91 cm$^3$

Hence 50 kg, i.e. 50 000 g has a volume of $\dfrac{50\ 000}{8.91}$ cm$^3$ = **5 612 cm$^3$**

(b) Volume of wire = area of circular cross-section $\times$ length of wire.
Hence 5612 cm$^3$ = area $\times$ (500 $\times$ 100 cm),

from which, area = $\dfrac{5612}{500 \times 100}$ cm$^2$ = **0.1122 cm$^2$**

(c) Area of circle = $\pi r^2$ or $\dfrac{\pi d^2}{4}$

Hence $0.1122 = \dfrac{\pi d^2}{4}$, from which $d = \sqrt{\left(\dfrac{4 \times 0.1122}{\pi}\right)} = 0.3780$ cm,

i.e. **diameter of cross-section is 3.780 mm**

---

*Problem 26* A boiler consists of a cylindrical section of length 8 m and diameter 6 m, on one end of which is surmounted a hemispherical section of diameter 6 m, and on the other end a conical section of height 4 m. Calculate the volume of the boiler and the total surface area.

The boiler is shown in *Fig 22.*

Volume of hemisphere,      $P = \frac{2}{3}\pi r^3 = \frac{2}{3} \times \pi \times 3^3 = 18\pi$ m$^3$

Volume of cylinder,      $Q = \pi r^2 h = \pi \times 3^2 \times 8 = 72\pi$ m$^3$

Volume of cone,      $R = \frac{1}{3}\pi r^2 h = \frac{1}{3} \times \pi \times 3^2 \times 4 = 12\pi$ m$^3$

**Total volume of boiler** $= 18\pi + 72\pi + 12\pi = 102\pi = $ **320.4 m$^3$**

145

Surface area of hemisphere

$$P = \frac{1}{2}(4\pi r^2) = 2 \times \pi \times 3^2 = 18\pi \text{ m}^2$$

Curved surface area of cylinder,

$$Q = 2\pi rh = 2 \times \pi \times 3 \times 8 = 48\pi \text{ m}^2$$

The slant height of the cone, $l$, is obtained by Pythagoras' theorem on triangle ABC.

$$l = \sqrt{(4^2+3^2)} = 5$$

Curved surface area of cone,

$$R = \pi rl = \pi \times 3 \times 5 = 15\pi \text{ m}^2$$

**Total surface area of boiler**

$$= 18\pi + 48\pi + 15\pi = 81\pi = 254.5 \text{ m}^2$$

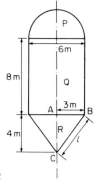

**Fig 22**

*Further problems on volumes and surface areas of regular solids may be found in section C following, problems 16 to 32, page 148.*

## C. FURTHER PROBLEMS ON AREAS AND VOLUMES

*Areas of plane figures*

1   A rectangular plate is 85 mm long and 42 mm wide. Find its area in square centimetres.

[35.7 cm$^2$]

2   A rectangular field has an area of 1.2 hectares and a length of 150 m. Find (a) its width and (b) the length of a diagonal. (1 hectare = 10 000 m$^2$).

[(a) 80 m; (b) 170 m]

3   Determine the area of each of the angle iron sections shown in *Fig 23*

[(a) 29 cm$^2$; (b) 650 mm$^2$]

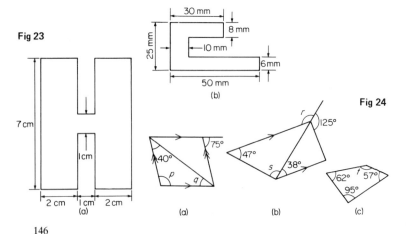

4   A rectangular garden measures 40 m by 15 m. A 1 m flower border is made round the two shorter sides and one long side. A circular swimming pool of diameter 8 m is constructed in the middle of the garden. Find, correct to the nearest square metre, the area remaining.

[482 m$^2$]

5   The area of a trapezium is 13.5 cm$^2$ and the perpendicular distance between its parallel sides is 3 cm. If the length of one of the parallel sides is 5.6 cm, find the length of the other parallel side.

[3.4 cm]

6   Find the angles $p$, $q$, $r$, $s$ and $t$ in Figs 24(a) to (c).

[$p = 105°$; $q = 35°$; $r = 142°$; $s = 95°$; $t = 146°$]

7   Name the types of quadrilateral shown in Figs 25(i) to (iv), and determine (a) the area, and (b) the perimeter of each.

⌈ (i) rhombus (a) 14 cm$^2$; (b) 16 cm;
│ (ii) parallelogram (a) 180 mm$^2$; (b) 80 mm;
│ (iii) rectangle (a) 3600 mm$^2$; (b) 300 mm;
⌊ (iv) trapezium (a) 190 cm$^2$; (b) 62.91 cm ⌋

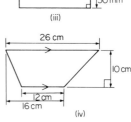

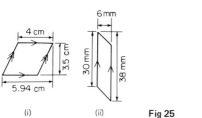

(i)          (ii)          **Fig 25**          (iv)

8   Determine the area of circles having (a) a radius of 4 cm; (b) a diameter of 30 mm; (c) a circumference of 200 mm.

[(a) 50.27 cm$^2$; (b) 706.9 mm$^2$; (c) 3183 mm$^2$]

9   An annulus has an outside diameter of 60 mm and an inside diameter of 20 mm. Determine its area.

[2513 mm$^2$]

10  If the area of a circle is 320 mm$^2$, find (a) its diameter, and (b) its circumference.

[(a) 20.19 mm; (b) 63.41 mm]

11  Calculate the areas of the following sectors of circles: (a) radius 9 cm, angle subtended at centre 75°; (b) diameter 35 mm, angle subtended at centre 48° 37′; (c) diameter 5 cm, angle subtended at centre 2.19 radians.

[(a) 53.01 cm$^2$; (b) 129.9 mm$^2$; (c) 6.84 cm$^2$]

12  Calculate the area of a regular octagon if each side is 20 mm and the width across the flats is 48.3 mm.

[1932 mm$^2$]

13  Determine the area of a regular hexagon which has sides 25 mm.

[1624 mm$^2$]

14  Find the area of triangle ABC if $\angle B = 90°$, $a = 4.87$ cm and $b = 7.54$ cm.

[14.02 cm$^2$]

15  The area of a park on a map is 500 mm$^2$. If the scale of the map is 1 to 40 000 determine the true area of the park in hectares (1 hectare = $10^4$ m$^2$).

[80 ha]

*Volumes and surface areas of regular solids*

16 A rectangular block of metal has dimensions of 40 mm by 25 mm by 15 mm. Determine its volume. Find also its mass if the metal has a density of 9 g/cm$^3$.

[15 cm$^3$; 135 g]

17 Determine the maximum capacity, in litres, of a fish tank measuring 50 cm by 40 cm by 2.5 m. (1 litre = 1000 cm$^3$.)

[500 l]

18 Determine how many cubic metres of concrete are required for a 120 m long path, 150 mm wide and 80 mm deep.

[1.44 m$^3$]

19 Calculate the volume of a metal tube whose outside diameter is 8 cm and whose inside diameter is 6 cm, if the length of the tube is 4 m.

[8 796 cm$^3$]

20 The volume of a cylinder is 400 cm$^3$. If its radius is 5.20 cm, find its height. Determine also its curved surface area.

[4.709 cm; 153.9 cm$^2$]

21 The diameter of two spherical bearings are in the ratio 2 : 5. What is the ratio of their volumes?

[8 : 125]

22 An engineering component has a mass of 400 g. If each of its dimensions are reduced by 30% determine its new mass.

[137.2 g]

23 If a cone has a diameter of 80 mm and a perpendicular height of 120 mm calculate its volume in cm$^3$ and its curved surface area.

[201.1 cm$^3$; 159.0 cm$^2$]

24 A cylinder is cast from a rectangular piece of alloy 5 cm by 7 cm by 12 cm. If the length of the cylinder is to be 60 cm, find its diameter.

[2.99 cm]

25 Find the volume and the total surface area of a regular hexagonal bar of metal of length 3 m if each side of the hexagon is 6 cm.

[28.060 cm$^3$; 1.099 m$^2$]

26 A square pyramid has a perpendicular height of 4 cm. If a side of the base is 2.4 cm long find the volume and total surface area of the pyramid.

[7.68 cm$^3$; 25.81 cm$^2$]

27 A sphere has a diameter of 6 cm. Determine its volume and surface area.

[113.1 cm$^3$; 113.1 cm$^2$]

28 Find the total surface area of a hemisphere of diameter 50 mm.

[5890 mm$^2$ or 58.90 cm$^2$]

29 Determine the mass of a hemispherical copper container whose external and internal radii are 12 cm and 10 cm. Assume that 1 cm$^3$ of copper weights 8.9 g.

[13.57 kg]

30 If the volume of a sphere is 566 cm$^3$, find its radius.

[5.131 cm]

31 A metal plumb bob comprises a hemisphere surmounted by a cone. If the diameter of the hemisphere and cone are each 4 cm and the total length is 5 cm, find its total volume.

[29.32 cm$^3$]

32 A marquee is in the form of a cylinder surmounted by a cone. The total height is 6 m and the cylindrical portion has a height of 3.5 m, with a diameter of 15 m. Calculate the surface area of material needed to make the marquee assuming 12% of the material is wasted in the process.        [393.4 m$^2$]

# 11 An introduction to trigonometry

## A. MAIN POINTS CONCERNED WITH TRIGONOMETRY

1  **Trigonometry** is the branch of mathematics which deals with the measurement of sides and angles of triangles, and their relationships with each other.

2  For the right-angled triangle shown in *Fig 1*, the three most important **trigonometrical ratios of acute angles** are, by definition:

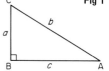

Fig 1

(i)  sine $A$ $= \dfrac{\text{opposite side}}{\text{hypotenuse}} = \dfrac{BC}{AC}$ , i.e. $\sin A = \dfrac{a}{b}$

(ii)  cosine $A = \dfrac{\text{adjacent side}}{\text{hypotenuse}} = \dfrac{AB}{AC}$ , i.e. $\cos A = \dfrac{c}{b}$

(iii)  tangent $A = \dfrac{\text{opposite side}}{\text{adjacent side}} = \dfrac{BC}{AB}$ , i.e. $\tan A = \dfrac{a}{c}$

3  **4-figure trigonometrical tables** give the values of the sine, cosine and tangent of any angle between $0°$ and $90°$, correct to the nearest minute, and are read in a similar manner to that used for squares, square roots and reciprocals (see chapter 3).

(i)  From **natural sine tables**, the value of a sine is seen to increase from 0 at $0°$ to a maximum of 1 at $90°$. For example,

$\sin 21°$ $= 0.3584$, $\quad \sin 21° \, 36' = 0.3681$,
$\sin 21° \, 38' = 0.3681 + 5$ (from the mean difference column equivalent to $2'$)

i.e. $\sin 21° \, 38' = 0.3686$.

Similarly, $\sin 47° \, 41' = 0.7395$ and $\sin 82° \, 9' = 0.9906$.

To find the angle whose sine is 0.6329, find in the table the nearest number **less than** 0.6329; in this case it is 0.6320, corresponding to $39° \, 12'$. 0.6320 is 9 less than 0.6329 and 9 in the mean difference column corresponds to $4'$. Hence the angle whose sine is 0.6329 is $39° \, 12' + 4'$, i.e. $39° \, 16'$.

Hence, arcsin $0.6329 = 39° \, 16'$.

('arcsin $\theta$'is a short way of writing "the angle whose sine is equal to $\theta$").

Similarly, arcsin $0.0875 = 5° \, 1'$ and arcsin $0.9146 = 66° \, 9'$.

(ii)  From **natural cosine tables**, the value of a cosine is seen to decrease from a maximum of 1 at $0°$ to 0 at $90°$. For example,

$\cos 37°$ $= 0.7986$
$\cos 37° \, 12' = 0.7965$
$\cos 37° \, 16' = 0.7965 - 7$ (from the mean difference column equivalent to $4'$)

i.e. $\cos 37° \, 16' = 0.7958$.

(Note, for cosines, the numbers in the difference columns are subtracted, not added.)

Similarly, cos 11° 53′ = 0.9786 and cos 84° 20′ = 0.0987.

To find the angle whose cosine is 0.5843, find in the table the nearest number **greater than** 0.5843; in this case it is 0.5850, which corresponds to 54° 12′. 0.5843 is 7 less than 0.5850, and 7 in the mean difference column corresponds to 3′. Hence arccos 0.5843 = 54° 12′+3′, i.e. 54° 15′.

Similarly, arccos 0.9551 = 17° 14′ and arccos 0.2490 = 75° 35′.

(iii) From **natural tangent tables**, the value of a tangent is seen to increase from 0 at 0° to 1 at 45° and then to infinity at 90°. Natural tangent tables are read in the same way as natural sine tables. For example, tan 32° 52′ = 0.6461 and tan 76° 14′ = 4.0820. Also, arctan 0.8729 = 41° 7′ and arctan 2.4862 = 68° 5′.

4   **Fractional and surd forms of trigonometric ratios.** In *Fig 2*, ABC is an equilateral triangle of side 2 units. AD bisects angle $A$ and bisects the side BC. Using Pythagoras' theorem on triangle ABD gives: AD = $\sqrt{(2^2-1^2)} = \sqrt{3}$.

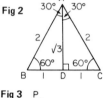

**Fig 2**

Hence sin 30° = $\dfrac{BD}{AB} = \dfrac{1}{2}$; cos 30° = $\dfrac{AD}{AB} = \dfrac{\sqrt{3}}{2}$;

$$\tan 30° = \dfrac{BD}{AD} = \dfrac{1}{\sqrt{3}}$$

sin 60° = $\dfrac{AD}{AB} = \dfrac{\sqrt{3}}{2}$; cos 60° = $\dfrac{BD}{AB} = \dfrac{1}{2}$;

$$\tan 60° = \dfrac{AD}{BD} = \sqrt{3}$$

**Fig 3**

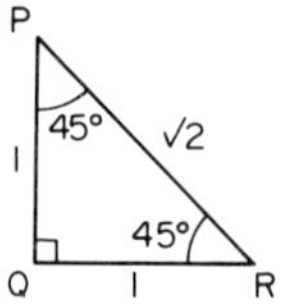

In *Fig 3*, PQR is an isosceles triangle with PQ = QR = 1 unit. By Pythagoras' theorem, PR = $\sqrt{(1^2+1^2)} = \sqrt{2}$.

Hence sin 45° = $\dfrac{1}{\sqrt{2}}$; cos 45° = $\dfrac{1}{\sqrt{2}}$; tan 45° = 1

A quantity which is not exactly expressible as a rational number is called a **surd**. For example, $\sqrt{2}$ and $\sqrt{3}$ are called surds because they cannot be expressed as a fraction and the decimal part may be continued indefinitely. For example, $\sqrt{2} = 1.4142135\ldots\ldots$

5   From paragraph 4, sin 30° = cos 60°, sin 45° = 45° and sin 60° = cos 30°.
In general, **sin $\theta$ = cos(90°−$\theta$)**
and        **cos $\theta$ = sin(90°−$\theta$)**
For example, it may be checked from tables that sin 25° = cos 65°, sin 42° = cos 48°, cos 84° 10′ = sin 5° 50′, and so on.

6   To 'solve a right-angled triangle' means 'to find the unknown sides and angles'. This is achieved by using (i) the theorem of Pythagoras; and/or (ii) trigonometric ratios.

7   If logarithms are used in calculations involving trigonometric ratios then tables of **'logarithms of sines'**, **'logarithms of cosines'**, and **'logarithms of tangents'** are useful aids. The method of reading such tables is the same as for natural sines, cosines and tangents (see paragraph 3). However, the proficient use of a **scientific notation calculator** is encouraged as an accurate and rapid aid for calculations.

8   If, in *Fig 4*, BC represents horizontal ground and AB a vertical flagpole, then the **angle of elevation** of the top of the flagpole, A, from the point C is the angle that the imaginary straight line AC must be raised (or elevated) from the horizontal CB, i.e. angle $\theta$.

**Fig 4**

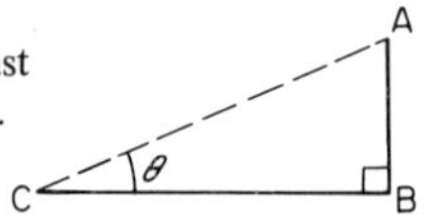

9  If, in *Fig 5*, PQ represents a vertical cliff and R a ship at sea, then the **angle of depression** of the ship from point P is the angle through which the imaginary straight line PR must be lowered (or depressed) from the horizontal to the ship, i.e. angle $\phi$. (Note, $\angle$PRQ is also $\phi$ — alternate angles between parallel lines.)

10  In *Fig 6*, let OR be a vector, 1 unit long and free to rotate anticlockwise about 0. In one revolution a circle is produced and is shown with 15° sectors. Each radius arm has a vertical and a horizontal component. For example, at 30°, the vertical component is TS and the horizontal component is OS.

*From trigonometric ratios,*

$\sin 30° = \dfrac{TS}{OT} = \dfrac{TS}{1}$ i.e. TS = $\sin 30°$ and

$\cos 30° = \dfrac{OS}{OT} = \dfrac{OS}{1}$ i.e. OS = $\cos 30°$

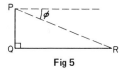

**Fig 5**

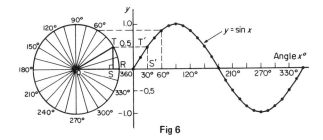

**Fig 6**

The vertical component TS may be projected across to T'S', which is the corresponding value of 30° on the graph of $y$ against angle $x°$. If all such vertical components as TS are projected on to the graph, then a **sine wave** is produced as shown in *Fig 6*. If all horizontal components such as OS are projected on to a graph of $y$ against angle $x°$, then a **cosine wave** is produced. It is easier to visualise these projections by redrawing the circle with the radius arm OR initially in a vertical position as shown in *Fig 7*. From *Figs 6 and 7* it is seen that a cosine curve is of the same form as the sine curve but is displaced by 90° (or $\pi/2$ radians).

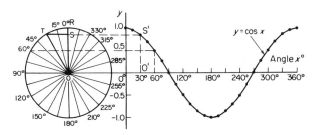

**Fig 7**

## B. WORKED PROBLEMS ON TRIGONOMETRY

*Problem 1*  Sketch a right-angled triangle ABC such that $\angle B = 90°$, AB = 5 cm and BC = 12 cm. Determine the length of AC and hence evaluate sin $C$, cos $C$ and tan $C$

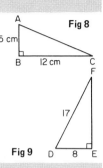

Fig 8

Fig 9

Triangle ABC is shown in *Fig 8*.
By Pythagoras' theorem, AC = $\sqrt{(5^2 + 12^2)}$ = 13

By definition, sin $C = \dfrac{\text{opposite side}}{\text{hypotenuse}}$ $= \dfrac{5}{13}$ or 0.3846

$\cos C = \dfrac{\text{adjacent side}}{\text{hypotenuse}}$ $= \dfrac{12}{13}$ or 0.9231

$\tan C = \dfrac{\text{opposite side}}{\text{adjacent side}}$ $= \dfrac{5}{12}$ or 0.4167

*Problem 2*  From *Fig 9*, find sin $D$, cos $D$ and tan $F$

By Pythagoras' theorem, $17^2 = 8^2 + EF^2$, from which, EF = $\sqrt{(17^2 - 8^2)}$ = 15

$\sin D = \dfrac{EF}{DF}$ $\dfrac{15}{17}$ or 0.8824;        $\cos D = \dfrac{DE}{DF}$ $\dfrac{8}{17}$ or 0.4706;

$\tan F = \dfrac{DE}{EF}$ $\dfrac{8}{15}$ or 0.5333

*Problem 3*  If $\cos X = \dfrac{24}{25}$, find sin $X$ and tan $X$

Fig 10

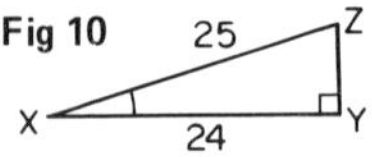

*Fig 10* shows a right-angled triangle XYZ with $\cos X = \dfrac{24}{25}$.
Using Pythagoras' theorem, $25^2 = 24^2 + YZ^2$, from which YZ = $\sqrt{(25^2 - 24^2)}$ = 7
Hence sin $X = \dfrac{YZ}{XZ}$ $\dfrac{7}{25}$ or 0.2800;  tan $X = \dfrac{YZ}{XY}$ $\dfrac{7}{24}$ or 0.2917

*Problem 4*  Use 4 figure tables to evaluate (a) sin 24° 52′; (b) cos 39° 35′; (c) tan 58° 45′

(a) sin 24° 48′ = 0.4195
    sin 24° 52′ = 0.4195+11 (from the mean difference column equivalent to 4′)
i.e. **sin 24° 52′ = 0.4206**

(b) cos 39° 30' = 0.7716
   cos 39° 35' = 0.7716−9 (from the mean difference column equivalent to 5')
i.e. **cos 39° 35' = 0.7707**
(c) tan 58° 42' = 1.6447
   tan 58° 45' = 1.6447+32 (from the mean difference column equivalent to 3')
i.e. **tan 58° 45' = 1.6479**

*Problem 5* From 4 figure tables find (a) arcsin 0.7844; (b) arccos 0.1262; (c) arctan 3.1614

(a) The nearest number less than 0.7844 in natural sine tables is 0.7837, corresponding to 51° 36'. The difference between 0.7844 and 0.7837 is 7. From the mean difference column 7 corresponds to 4'.
   Hence arcsin 0.7844 = 51° 36'+4' = **51° 40'**.
(b) The nearest number greater than 0.1262 in natural cosine tables is 0.1271 corresponding to 82° 42'. The difference between 0.1262 and 0.1271 is 9. From the mean difference column 9 corresponds to 3'.
   Hence arccos 0.1262 = 82° 42'+3' = **82° 45'**.
(c) The nearest number less than 3.1614 in natural tangent tables is 3.1524, corresponding to 72° 24'. The difference between 3.1614 and 3.1524 is 90. From the mean difference column 90 corresponds closest to 3'.
   Hence acrtan 3.1614 = 72° 24'+3' = **72° 27'**.

*Problem 6* Evaluate $\dfrac{4.2 \tan 49° \ 26'-3.7 \sin 66° \ 1'}{7.1 \cos 29° \ 34'}$ , correct to 3 significant figures.

From 4-figure tables, tan 49° 26' = 1.1681, sin 66° 1' = 0.9136 and cos 29° 34' = 0.8698.

Hence $\dfrac{4.2 \tan 49° \ 26'-3.7 \sin 66° \ 1'}{7.1 \cos 29° \ 34'}$ $= \dfrac{(4.2 \times 1.1681)-(3.7 \times 0.9136)}{(7.1 \times 0.8698)}$

$= \dfrac{4.9060-3.3803}{6.1756}$

$= \dfrac{1.5257}{6.1756} = 0.2471 =$ **0.247, correct to 3 significant figures**

*Problem 7* Using surd forms, evaluate $\dfrac{3 \tan 60°-2 \cos 30°}{\tan 30°}$

From paragraph 4, $\tan 60° = \sqrt{3}$, $\cos 30° = \dfrac{\sqrt{3}}{2}$ and $\tan 30° = \dfrac{1}{\sqrt{3}}$

Hence $\dfrac{3 \tan 60°-2 \cos 30°}{\tan 30°} = \dfrac{3(\sqrt{3})-2\left(\dfrac{\sqrt{3}}{2}\right)}{\dfrac{1}{\sqrt{3}}} = \dfrac{3\sqrt{3}-\sqrt{3}}{\dfrac{1}{\sqrt{3}}}$

153

$$= \frac{2\sqrt{3}}{\frac{1}{\sqrt{3}}} = 2\sqrt{3}\left(\frac{\sqrt{3}}{1}\right) = 2(3) = 6$$

*Problem 8* Using tables of common logarithms and 'logarithms of sines' to evaluate angle $X$ given:
$$\frac{4.2}{\sin 35°} = \frac{6.5}{\sin X} .$$

Rearranging gives: $\sin X = \dfrac{6.5 \sin 35°}{4.2}$

$$X = \arcsin \frac{6.5 \sin 35°}{4.2}$$

Hence $X = \mathbf{62° \ 36'}$

| Number | Logarithms |
|---|---|
| 6.5 | 0.8129 |
| log sin 35° | $\overline{1}.7586$ |
| | 0.5715 |
| 4.2 | 0.6232 |
| 62° 36′ | $\overline{1}.9483$ |

Note that the angle corresponding to the logarithm $\overline{1}.9483$ is read directly from 'logarithms of sines' tables.

*Problem 9* In triangle PQR shown in *Fig 11*, find the lengths of PQ and PR.

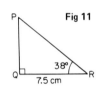

Fig 11

$\tan 38° = \dfrac{PQ}{QR} = \dfrac{PQ}{7.5} .$

Hence PQ = 7.5 tan 38° = 7.5 (0.7813) = **5.860 cm**

$\cos 38° = \dfrac{QR}{PR} = \dfrac{7.5}{PR} .$

Hence PR = $\dfrac{7.5}{\cos 38°} = \dfrac{7.5}{0.7880}$ = **9.518 cm**

(Check: Using Pythagoras' theorem $(7.5)^2 + (5.860)^2 = 90.59 = (9.518)^2$)

*Problem 10* Solve the triangle ABC shown in *Fig 12*.

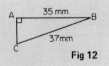

Fig 12

To 'solve triangle ABC' means 'to find the length AC and angles $B$ and $C$'

$\sin C = \dfrac{35}{37} = 0.9459.$ Hence $C = \arcsin 0.9459 = \mathbf{71° \ 4'}$

$B = 180° - 90° - 71° \ 4' = \mathbf{18° \ 56'}$ (since angles in a triangle add up to $180°$)

$\sin B = \dfrac{AC}{37}$ hence AC = 37 sin 18° 56′ = 37(0.3245) = **12.0 mm**

(Check: Using Pythagoras' theorem $37^2 = 35^2 + 12^2$)

*Problem 11* Solve triangle XYZ given $\angle X = 90°$, $\angle Y = 23°\ 17'$ and YZ = 20.0 mm. Determine also its area.

It is always advisable to make a reasonably accurate sketch so as to visualise the expected magnitudes of unknown sides and angles. Such a sketch is shown in *Fig 13*.

**Fig 13**

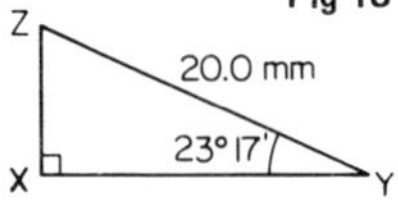

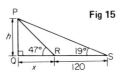

$\angle Z = 180° - 90° - 23°17' = \textbf{66° 43'}$.

$\sin 23°\ 17' = \dfrac{XZ}{20.0}$. Hence $XZ = 20.0 \sin 23°\ 17'$
$= 20.0(0.3953) = \textbf{7.906 mm}$

$\cos 23°\ 17' = \dfrac{XY}{20.0}$. Hence $XY = 20.0 \cos 23°\ 17' = 20.0(0.9185) = \textbf{18.37 mm}$.

(Check: Using Pythagoras' theorem $(18.37)^2 + (7.906)^2 = 400.0 = (20.0)^2$).

Area of triangle $XYZ = \dfrac{1}{2}$(base)(perpendicular height) $= \dfrac{1}{2}(XY)(XZ)$

$= \dfrac{1}{2}(18.37)(7.906)$

$= \textbf{72.62 mm}^2$

*Problem 12* An electricity pylon stands on horizontal ground. At a point 80 m from the base of the pylon, the angle of elevation of the top of the pylon is 23°. Calculate the height of the pylon to the nearest metre.

*Fig 14* shows the pylon AB and the angle of elevation of A from point C is 23°.

**Fig 14**

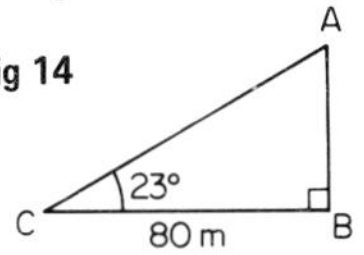

$\tan 23° = \dfrac{AB}{BC} = \dfrac{AB}{80}$

Hence height of pylon $AB = 80 \tan 23°$

$= 80(0.4245) = 33.96$ m

$= \textbf{34 m to the nearest metre}$

*Problem 13* A surveyor measures the angle of elevation of the top of a perpendicular building as 19°. He moves 120 m nearer the building and finds the angle of elevation is now 47°. Determine the height of the building.

The building PQ and the angles of elevation are shown in *Fig 15*.

**Fig 15**

In triangle PQS, $\tan 19° = \dfrac{h}{x+120}$;

hence $h = \tan 19°\ (x+120)$

i.e. $h = 0.3443(x+120)$    (1)

In triangle PQR, $\tan 47° = \dfrac{h}{x}$;

hence $h = \tan 47°\ (x)$

i.e. $h = 1.0724x$    (2)

155

Equating equations (1) and (2) gives: $\quad 0.3443(x+120) = 1.0724x$

$$0.3443x+(0.3443)(120) = 1.0724x$$
$$(0.3443)(120) = (1.0724-0.3443)x$$
$$41.316 = 0.7281x$$
$$x = \frac{41.316}{0.7281} = 56.74 \text{ m}$$

From equation (2), height of building $h = 1.0724x = 1.0724(56.74) = \textbf{60.85 m}$

*Problem 14* The angle of depression of a ship viewed at a particular instant from the top of a 75 m vertical cliff is 30°. Find the distance of the ship from the base of the cliff at this instant. The ship is sailing away from the cliff at constant speed and 1 minute later its angle of depression from the top of the cliff is 20°. Determine the speed of the ship in km/h.

*Fig 16* shows the cliff AB, the initial position of the ship at C and the final position at D. Since the angle of depression is initially 30° then $\angle ACB = 30°$ (alternate angles between parallel lines).

$$\tan 30° = \frac{AB}{BC} = \frac{75}{BC}.$$

Hence $BC = \dfrac{75}{\tan 30°} = \dfrac{75}{0.5774}$

$= \textbf{129.9 m} = $ **initial position of ship**

In triangle ABD,

$$\tan 20° = \frac{AB}{BD} = \frac{75}{BC+CD} = \frac{75}{129.9+x}$$

**Fig 16**

Hence $129.9+x = \dfrac{75}{\tan 20°} = \dfrac{75}{0.3640} = 206.0$ m,

from which $x = 206.0-129.9$
$= 76.1$ m

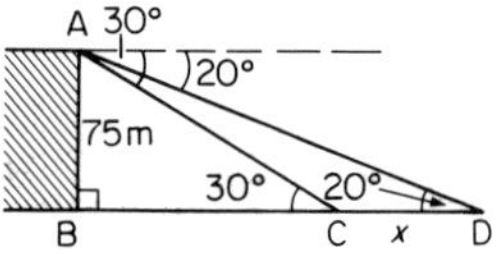

Thus the ship sails 76.1 m in 1 minute, i.e. 60 s.

Hence speed of ship $= \dfrac{\text{distance}}{\text{time}} = \dfrac{76.1}{60}$ m/s $= \dfrac{76.1 \times 60 \times 60}{60 \times 1000}$ km/h $= \textbf{4.566 km/h}$

*Further problems on trigonometry may be found in section C following, problems 1 to 20.*

## C. FURTHER PROBLEMS ON TRIGONOMETRY

1  Sketch a triangle XYZ such that $\angle Y = 90°$, $XY = 9$ cm and $YZ = 40$ cm. Determine $\sin Z$, $\cos Z$, $\tan X$ and $\cos X$.

$$\left[\sin Z = \frac{9}{41}; \cos Z = \frac{40}{41}; \tan X = \frac{40}{9}; \cos X = \frac{9}{41}\right]$$

**Fig 17**

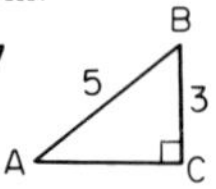

2  In triangle ABC shown in *Fig 17*, find $\sin A$, $\cos A$, $\tan A$, $\sin B$, $\cos B$ and $\tan B$.

$$\left[\sin A = \frac{3}{5}, \cos A = \frac{4}{5}, \tan A = \frac{3}{4}; \sin B = \frac{4}{5}; \cos B = \frac{3}{5}; \tan B = \frac{4}{3}\right]$$

3  If $\cos A = \dfrac{15}{17}$ find $\sin A$ and $\tan A$, in fraction form.

$$\left[\sin A = \frac{8}{17}; \tan A = \frac{8}{15}\right]$$

4  If $\tan X = \dfrac{15}{112}$, find $\sin X$ and $\cos X$, in fraction form.

$$\left[\sin X = \frac{15}{113}; \cos X = \frac{112}{113}\right]$$

5  Use 4 figure tables to find the values of: (a) $\sin 15°\ 12'$; (b) $\sin 34°\ 51'$;
(c) $\sin 78°\ 11'$; (d) $\cos 43°\ 17'$; (e) $\cos 65°\ 45'$; (f) $\tan 29°\ 17'$; (g) $\tan 68°\ 38'$
[(a) 0.2622; (b) 0.5714; (c) 0.9788; (d) 0.7280; (e) 0.4107; (f) 0.5608; (g) 2.5560]

6  From 4 figure tables determine (a) arcsin 0.2678, (b) arccos 0.6752; (c) arctan 1.4786.

[(a) 15° 32'; (b) 47° 32'; (c) 55° 56']

7  Evaluate the following, each correct to 4 significant figures:

(a) $4 \cos 56°\ 19' - 3 \sin 21°\ 57'$;  (b) $\dfrac{11.5 \tan 49°\ 11' - \sin 90°}{3 \cos 45°}$

(c) $\dfrac{5 \sin 86°\ 3'}{3 \tan 14°\ 29' - 2 \cos 31°\ 9'}$

[(a) 1.097; (b) 5.805; (c) −5.325]

8  Evaluate the following without using tables or calculators, leaving, where necessary, in surd form:

(a) $3 \sin 30° - 2 \cos 60°$            (b)  $5 \tan 60° - 3 \sin 60°$

(c) $\dfrac{\tan 60°}{3 \tan 30°}$            (d)  $(\tan 45°)(4 \cos 60° - 2 \sin 60°)$

(e) $\dfrac{\tan 60° - \tan 30°}{1 + \tan 30° \tan 60°}$

$$\left[\text{(a) } \frac{1}{2}; \text{(b) } \frac{7}{2}\sqrt{3}; \text{(c) } 1; \text{(d) } 2 - \sqrt{3}; \text{(e) } \frac{1}{\sqrt{3}}\right]$$

9  Use 4 figure tables of logarithms of trigonometric ratios to evaluate the following:

(a) arcsin $\dfrac{4.32 \sin 42°\ 16'}{7.86}$            (b)  $\dfrac{(\sin 34°\ 27')(\cos 69°\ 2')}{(2 \tan 53°\ 39')}$

[(a) 21° 42' (b) 0.0745]

10 Solve the triangles shown in *Fig 18*.

(i)

(ii)

(iii)

**Fig 18**

$$\left[\begin{array}{l} \text{(i) BC} = 3.50 \text{ cm}; \text{AB} = 6.10 \text{ cm}; \angle B = 55° \\ \text{(ii) FE} = 5 \text{ cm}; \angle E = 53°\ 8'; \angle F = 36°\ 52' \\ \text{(iii) GH} = 9.841 \text{ mm}; \text{GI} = 11.32 \text{ mm}; \angle H = 49° \end{array}\right]$$

11 Solve the triangles shown in *Fig 19* and find their areas.

$$\left[\begin{array}{l} \text{(i) KL} = 5.43 \text{ cm}, \text{JL} = 8.62 \text{ cm}, \angle J = 39°, \text{Area} = 18.19 \text{ cm}^2 \\ \text{(ii) MN} = 28.86 \text{ mm}, \text{NO} = 13.82 \text{ mm}, \angle O = 64° \ 25', \text{Area} = 199.4 \text{ mm}^2 \\ \text{(iii) PR} = 7.934 \text{ m}, \angle Q = 65° \ 3', \angle R = 24° \ 57', \text{Area} = 14.64 \text{ m}^2 \end{array}\right]$$

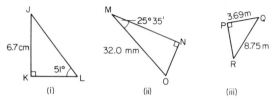

**Fig 19**

12 A ladder rests against the top of the perpendicular wall of a building and makes an angle of 67° with the ground. If the foot of the ladder is 12 m from the wall, calculate the height of the building.

[28.27 m]

13 A vertical tower stands on level ground. At a point 105 m from the foot of the tower the angle of elevation of the top is 19°. Find the height of the tower.

[36.15 m]

14 If the angle of elevation of the top of a vertical 30 m high aerial is 32°, how far is it to the aerial?

[48 m]

15 From the top of a vertical cliff 90 m high the angle of depression of a boat is 19° 50'. Determine the distance of the boat from the cliff.

[249.5 m]

16 From a point on horizontal ground a surveyor measures the angle of elevation of the top of a flagpole as 18° 40'. He moves 50 m nearer to the flagpole and measures the angle of elevation as 26° 22'. Determine the height of the flagpole.

[53.0 m]

17 A flagpole stands on the edge of the top of a building. At a point 200 m from the building the angles of elevation of the top and bottom of the pole are 32° and 30° respectively. Calculate the height of the flagpole.

[9.50 m]

18 From a ship at sea, the angles of elevation of the top and bottom of a vertical lighthouse standing on the edge of a vertical cliff are 31° and 26° respectively. If the light house is 25.0 m high, calculate the height of the cliff.

[107.8 m]

19 From a window 4.2 m above horizontal ground the angle of depression of the foot of a building across the road is 24° and the angle of elevation of the top of the same building is 34°. Determine, correct to the nearest centimetre, the width of the road and the height of the building.

[9.43 m; 10.56 m]

20 The elevation of a tower from two points, one due east of the tower and the other due west of it are 20° and 24° respectively and the two points of observation are 300 m apart. Find the height of the tower to the nearest metre.

[60 m]

# 12 An introduction to statistics

## A. MAIN POINTS CONCERNED WITH COLLECTING, REPRESENTING AND INTERPRETING STATISTICAL DATA

1 Data is obtained largely by two methods:
   (a) by counting − for example, the number of stamps sold by a post office in equal periods of time, and
   (b) by measurement − for example, the heights of a group of people.
2 When data is obtained by counting and only whole numbers are possible, the data is called **discrete**. Measured data can have any value within certain limits and is called **continuous**.
3 A **set** is a group of data and an individual value within the set is called a **member** of the set. Thus, if the masses of five people are measured correct to the nearest 0.1 kilogram and are found to be 53.1 kg, 59.4 kg, 62.1 kg, 77.8 kg and 64.4 kg, then the set of masses in kilograms for these five people is:

$$\{ 53.1, 59.4, 62.1, 77.8, 64.4 \}$$

and one of the members of the set is 59.4.

   A set containing all the members is called a **population**. Some members selected at random from a population is called a **sample**. Thus all car registration numbers form a population, but the registration numbers of, say, 20 cars taken at random throughout the country is a sample drawn from that population.
4 The number of times that the value of a member occurs in a set is called the **frequency** of that member. Thus in the set: $\{ 2, 3, 4, 5, 4, 2, 4, 7, 9 \}$, member 4 has a frequency of three, member 2 has a frequency of 2 and the other members have a frequency of one.
5 The **relative frequency** with which any member of a set occurs is given by the ratio:

$$\frac{\text{frequency of member}}{\text{total frequency of all members}}$$

For the set: $\{ 2, 3, 5, 4, 7, 5, 6, 2, 8 \}$, the relative frequency of member 5 is $\frac{2}{9}$. Often, relative frequency is expressed as a percentage and the **percentage relative frequency** is: (relative frequency $\times$ 100)%.
6 When the number of members in a set is comparatively small, say ten or less, the data can be represented diagrammatically without further analysis. For sets having more than ten members, those members having similar values are grouped together into **classes** to form a **frequency distribution**. To assist in accurately counting members in the various classes, a **tally diagram** is used, (see *problem 6*). A frequency distribution is merely a table showing classes and their corresponding frequencies (see *Problem 6*). The new set of values obtained by forming a frequency distribution is called **grouped data**.

7  A cumulative frequency distribution is a table showing the cumulative frequency for each value of upper class boundary (see *Fig 6*). The cumulative frequency for a particular value of upper class boundary is obtained by adding the frequency of the class to the sum of the previous frequencies. A cumulative frequency distribution is shown in *Problem 8*.

8  Ungrouped data can be presented diagrammatically in several ways and these include:
   (a) **pictograms**, in which pictorial symbols are used to represent quantities, (see *Problem 1*),
   (b) **horizontal bar charts**, having data represented by equally spaced horizontal rectangles, (see *Problem 2*), and
   (c) **vertical bar charts**, in which data is represented by equally spaced vertical rectangles, (see *Problem 3*).

9  Trends in ungrouped data over equal periods of time can be presented diagrammatically by a **percentage component bar chart**. In such a chart, equally spaced rectangles of any width, but whose height corresponds to 100%, are constructed. The rectangles are then subdivided into values corresponding to the percentage relative frequencies of the members, (see *Problem 4*).

10 A **pie diagram** is used to show diagrammatically the parts making up the whole. In a pie diagram, the area of a circle represents the whole, and the areas of the sectors of the circle are made proportional to the parts which make up the whole, (see *Problem 5*).

11 Grouped data may be presented diagrammatically in three ways:
   (a) by using a **frequency polygon**, which is the graph produced by plotting frequency against class mid-point values and joining the co-ordinates with straight lines, (see *Problem 6*),
   (b) by producing a **histogram**, in which the areas of vertical, adjacent rectangles are made proportional to the frequencies of the classes, (see *Problem 6*), and
   (c) by drawing an **ogive** or **cumulative frequency distribution curve**, which is a graph produced by plotting cumulative frequency against upper class boundary values and joining the co-ordinates by straight lines (see *Problem 8*).

## B.  WORKED PROBLEMS ON STATISTICS

*Problem 1*  The number of television sets repaired in a workshop by a technician in six, one-month periods is as shown below. Present this data as a pictogram.

| Month | January | February | March | April | May | June |
|---|---|---|---|---|---|---|
| Number repaired | 11 | 6 | 15 | 9 | 13 | 8 |

Each symbol shown in *Fig 1* represents two television sets repaired. Thus, in January, $5\frac{1}{2}$ symbols are used to represent the 11 sets repaired, in February, 3 symbols are used to represent the 6 sets repaired, and so on.

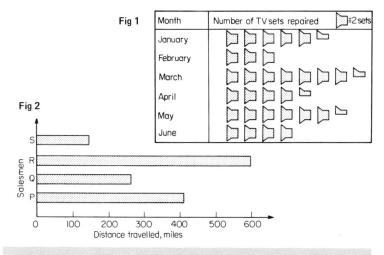

**Fig 1**

| Month | Number of TV sets repaired | ▭=2 sets |
|-------|---------------------------|----------|
| January | | |
| February | | |
| March | | |
| April | | |
| May | | |
| June | | |

**Fig 2**

Salesmen (P, Q, R, S) — Distance travelled, miles

---

*Problem 2* The distance in miles travelled by four salesmen in a week are as shown below.

| Salesmen | P | Q | R | S |
|----------|-----|-----|-----|-----|
| Distance travelled (miles) | 413 | 264 | 597 | 143 |

Use a horizontal bar chart to represent this data diagrammatically.

Equally spaced horizontal rectangles of any width, but whose length is proportional to the distance travelled, are used. Thus, the length of the rectangle for salesman P is proportional to 413 miles, and so on. The horizontal bar chart depicting this data is shown in *Fig 2*.

---

*Problem 3* The number of issues of tools or materials from a store in a factory is observed for seven, one-hour periods in a day, and the results of the survey are as follows:

| Period | 1 | 2 | 3 | 4 | 5 | 6 | 7 |
|--------|-----|-----|-----|-----|-----|-----|-----|
| Number of issues | 34 | 17 | 9 | 5 | 27 | 13 | 6 |

Present this data on a vertical bar chart.

In a vertical bar chart, equally spaced vertical rectangles of any width, but whose height is proportional to the quantity being represented, are used. Thus the height of the rectangle for period 1 is proportional to 34 units, and so on. The vertical bar chart depicting this data is shown in *Fig 3*.

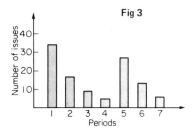

**Fig 3**

161

*Problem 4* The numbers of various types of dwellings sold by a company annually over a three-year period are as shown below. Draw percentage component bar charts to present this data.

|  | Year 1 | Year 2 | Year 3 |
|---|---|---|---|
| 4-roomed bungalows | 24 | 17 | 7 |
| 5-roomed bungalows | 38 | 71 | 118 |
| 4-roomed houses | 44 | 50 | 53 |
| 5-roomed houses | 64 | 82 | 147 |
| 6-roomed houses | 30 | 30 | 25 |

A table of percentage relative frequency values, correct to the nearest 1%, is the first requirement. Since,

$$\text{percentage relative frequency} = \frac{\text{frequency of member} \times 100}{\text{total frequency}}$$

then for 4-roomed bungalows in year 1:

$$\text{percentage relative frequency} = \frac{24 \times 100}{24 + 38 + 44 + 64 + 30} = 12\%$$

The percentage relative frequencies of the other types of dwellings for each of the three years are similarly calculated and the results are as shown in the table below.

|  | Year 1 | Year 2 | Year 3 |
|---|---|---|---|
| 4-roomed bungalows | 12% | 7% | 2% |
| 5-roomed bungalows | 19% | 28% | 34% |
| 4-roomed houses | 22% | 20% | 15% |
| 5-roomed houses | 32% | 33% | 42% |
| 6-roomed houses | 15% | 12% | 7% |

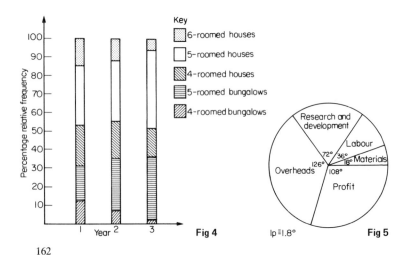

Key
▦ 6-roomed houses
☐ 5-roomed houses
▨ 4-roomed houses
▤ 5-roomed bungalows
▨ 4-roomed bungalows

Fig 4

Research and development / Labour / 72° / 36° / 18° Materials / 126° / Overheads / 108° / Profit

1p ≡ 1.8°          Fig 5

The percentage component bar chart is produced by constructing three equally spaced rectangles of any width, corresponding to the three years. The heights of the rectangles correspond to 100% relative frequency, and are sub-divided into the values in the table of percentages shown above. A key is used (different types of shading or different colour schemes) to indicate corresponding percentage values in the rows of the table of percentages. The percentage component bar chart is shown in *Fig 4*.

*Problem 5* The retail price of a product costing £2 is made up as follows: materials 10p, labour 20p, research and development 40p, overheads 70p, profit 60p. Present this data on a pie diagram.

A circle of any radius is drawn, and the area of the circle represents the whole, on which in this case is £2. The circle is sub-divided into sectors so that the areas of the sectors are proportional to the parts, i.e. the parts which make up the total retail price. For the area of a sector to be proportional to a part, the angle at the centre of the circle must be proportional to that part. The whole, £2 or 200p, corresponds to 360°. Therefore,

10p corresponds to $360 \times \dfrac{10}{200}$ degrees, i.e. 18°,

20p corresponds to $360 \times \dfrac{20}{200}$ degrees, i.e. 36°

and so on, giving the angles at the centre of the circle for the parts of the retail price as: 18°, 36°, 72°, 126° and 108° respectively.

The pie diagram is shown in *Fig 5*.

*Problem 6* The masses of 50 ingots in kilograms are measured correct to the nearest 0.1 kg and the results are as shown below. Produce a frequency distribution having about 7 classes for this data and then present the grouped data as (a) a frequency polygon and (b) a histogram.

| | | | | | | | | | |
|---|---|---|---|---|---|---|---|---|---|
| 8.0 | 8.6 | 8.2 | 7.5 | 8.0 | 9.1 | 8.5 | 7.6 | 8.2 | 7.8 |
| 8.3 | 7.1 | 8.1 | 8.3 | 8.7 | 7.8 | 8.7 | 8.5 | 8.4 | 8.5 |
| 7.7 | 8.4 | 7.9 | 8.8 | 7.2 | 8.1 | 7.8 | 8.2 | 7.7 | 7.5 |
| 8.1 | 7.4 | 8.8 | 8.0 | 8.4 | 8.5 | 8.1 | 7.3 | 9.0 | 8.6 |
| 7.4 | 8.2 | 8.4 | 7.7 | 8.3 | 8.2 | 7.9 | 8.5 | 7.9 | 8.0 |

The **range** of the data is the member having the largest value minus the member having the smallest value. Inspection of the set of data shows that:

range = 9.1−7.1 = 2.0

The size of each class is given approximately by $\dfrac{\text{range}}{\text{number of classes}}$.

Since about 7 classes are required, the size of each class is 2.0/7, that is approximately 0.3.

The terms used in connection with grouped data are shown in *Fig 6(a)*. The size of a class is given by the **upper class boundary** value minus the **lower class boundary** value, and in this problem the size is 0.3. To achieve about seven classes spanning the range of values from 7.1 to 9.1, the **class limits** are selected as 7.1 to 7.3, 7.4 to 7.6, 7.7 to 7.9, and so on. The **class mid-point** value is given by

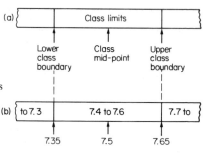

Fig 6

$$\frac{\text{upper class boundary value} - \text{lower class boundary value}}{2}$$

Thus the class mid-point for the 7.4 to 7.6 class is $(7.65 - 7.35)/2$, i.e. 7.5. The various values associated with the 7.4 to 7.6 class are shown in *Fig 6(b)*.

To assist with accurately determining the number in each class, a **tally diagram** is produced as shown in *Table 1(a)*. This is obtained by listing the classes in the left-hand column and then inspecting each of the 50 members of the set of data in turn and allocating it to the appropriate class by putting a '1' in the appropriate row. Each fifth '1' allocated to a particular row is marked as an oblique line to help with final counting.

A **frequency distribution** for the data is shown in *Table 1(b)* and lists classes and their corresponding frequencies. Class mid-points are also shown in this table, since they are used when constructing the frequency polygon and histogram.

A **frequency polygon** is shown in *Fig 7(a)*, the co-ordinates corresponding to the class mid-point/frequency values, given in *Table 1(b)*. The co-ordinates are joined by straight lines and the polygon is 'anchored-down' at each end by joining to the next class mid-point value and zero frequency.

A **histogram** is shown in *Fig 7(b)*, the width of a rectangle corresponding to (upper class boundary value − lower class boundary value) and height corresponding to the class frequency. The easiest way to draw a histogram is to mark class mid-point values on the horizontal scale and to draw the rectangles symmetrically about the appropriate class mid-point values and touching one another. A histogram for the data given in *Table 1(b)* is shown in *Fig 7(b)*.

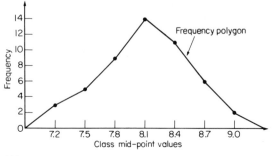

Fig 7(a)

| Class | Tally | | Class | Class mid-point | Frequency |
|-------|-------|---|-------|-----------|-----------|
| 7.1 to 7.3 | III | | 7.1 to 7.3 | 7.2 | 3 |
| 7.4 to 7.6 | ₩Ht | | 7.4 to 7.6 | 7.5 | 5 |
| 7.7 to 7.9 | ₩Ht IIII | | 7.7 to 7.9 | 7.8 | 9 |
| 8.0 to 8.2 | ₩Ht ₩Ht IIII | | 8.0 to 8.2 | 8.1 | 14 |
| 8.3 to 8.5 | ₩Ht ₩Ht I | | 8.3 to 8.5 | 8.4 | 11 |
| 8.6 to 8.8 | ₩Ht I | | 8.6 to 8.8 | 8.7 | 6 |
| 8.9 to 9.1 | II | | 8.9 to 9.1 | 9.0 | 2 |

(a) Tally diagram          (b) Frequency distribution

**Table 1**

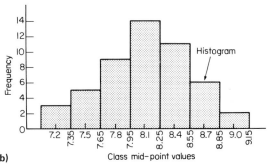

**Fig 7(b)**          Class mid-point values

Histogram

---

*Problem 7*

(a) Using the data given in *Fig 2* only, calculate the amount of money paid to each salesman for travelling expenses, if they are paid an allowance of 17.5p per mile.
(b) Using the data presented in *Fig 4*, comment on the housing trends over the three-year period.
(c) Determine the profit made by selling 700 units of the product shown in *Fig 5*.

---

(a) By measuring the length of rectangle P the mileage covered by salesman P is equivalent to 413 miles. Hence salesman P receives a travelling allowance of
$\frac{£413 \times 17.5}{100}$ , i.e. **£72.28**.
Similarly, for salesman Q, miles travelled are 264 and his allowance is
$\frac{£264 \times 17.5}{100}$ , i.e. **£46.20**.
Salesman R travels 597 miles and he receives
$\frac{£597 \times 17.5}{100}$ , i.e. **£104.48**.
Finally, salesman S receives
$\frac{£143 \times 17.5}{100}$ , that is, **£25.03**.

165

(b) An analysis of *Fig 4* shows that 5-roomed bungalows and 5-roomed houses are becoming more popular, the greatest change in the three years being a 15% increase in the sales of 5-roomed bungalows.

(c) Since 1.8° corresponds to 1p and the profit occupies 108° of the pie diagram, then the profit per unit is

$$\frac{108 \times 1}{1.8}$$ , that is, 60p.

The profit when selling 700 units of the product is

£ $\frac{700 \times 60}{100}$ , that is, **£420**.

---

*Problem 8* The frequency distribution for the masses in kilograms of 50 ingots is:

| | | |
|---|---|---|
| 7.1 to 7.3  3, | 7.4 to 7.6  5, | 7.7 to 7.9  9, |
| 8.0 to 8.2  14, | 8.3 to 8.5  11, | 8.6 to 8.8  6, |
| 8.9 to 9.1  2. | | |

Form a cumulative frequency distribution for this data and draw the corresponding ogive.

---

A cumulative frequency distribution is a table giving values of cumulative frequency for the values of upper class boundaries, and is shown in *Table 2*. Columns 1 and 2 show the classes and their frequencies. Column 3 lists the upper class boundary values for the classes given in column 1. Column 4 gives the cumulative frequency values for all frequencies less than the upper class boundary values given in column 3. Thus, for example, for the 7.7 to 7.9 class shown in row 3, the cumulative frequency value is the sum of all frequencies having values of less than 7.95, i.e., 3+5+9 = 17, and so on.

The ogive for the cumulative frequency distribution given in *Table 2* is shown in *Fig 8*. The co-ordinates corresponding to each upper class boundary/cumulative frequency value are plotted and the co-ordinates are joined by straight lines. (*Note*: not the best curve drawn through the co-ordinates as in experimental work.) The ogive is 'anchored' at its start by adding the co-ordinate (7.05, 0).

TABLE 2

| 1 | 2 | 3 | 4 |
|---|---|---|---|
| Class | Frequency | Upper class boundary | Cumulative frequency |
| | | Less than | |
| 7.1–7.3 | 3 | 7.35 | 3 |
| 7.4–7.6 | 5 | 7.65 | 8 |
| 7.7–7.9 | 9 | 7.95 | 17 |
| 8.0–8.2 | 14 | 8.25 | 31 |
| 8.3–8.5 | 11 | 8.55 | 42 |
| 8.6–8.8 | 6 | 8.85 | 48 |
| 8.9–9.1 | 2 | 9.15 | 50 |

*Further problems on statistics may be found in section C, following, problems 1 to 22*

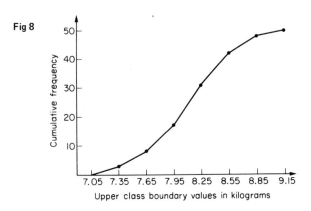

**Fig 8**

Cumulative frequency (y-axis): 10, 20, 30, 40, 50

Upper class boundary values in kilograms (x-axis): 7.05  7.35  7.65  7.95  8.25  8.55  8.85  9.15

## C. FURTHER PROBLEMS ON STATISTICS

1  The number of bottles of milk delivered by each of 5 milkmen in a one-hour period is as follows:

| Milkman | A | B | C | D | E |
|---|---|---|---|---|---|
| Number of bottles | 41 | 73 | 62 | 58 | 84 |

Present this data as a pictogram.

⎡If one symbol is used to represent 10 bottles, then, working correct to⎤
the nearest 5 bottles, the number of symbols are A 4, B $7\frac{1}{2}$, C 6, D 6 and
⎣E $8\frac{1}{2}$ .⎦

2  The number of vehicles passing a stationary observer on a road in six ten-minute intervals is as shown. Draw a pictogram to represent this data.

| Period of Time | 1 | 2 | 3 | 4 | 5 | 6 |
|---|---|---|---|---|---|---|
| Number of Vehicles | 35 | 44 | 62 | 68 | 49 | 41 |

⎡If one symbol is used to represent 10 vehicles, working correct to the⎤
⎣nearest 5 vehicles gives $3\frac{1}{2}$, $4\frac{1}{2}$, 6, 7, 5 and 4 symbols respectively.⎦

3  The number of components produced by a factory in a week is as shown below:

| Day | Mon | Tues | Wed | Thur | Fri |
|---|---|---|---|---|---|
| Number of Components | 1 580 | 2 190 | 1 840 | 2 385 | 1 280 |

Show this data on a pictogram.

⎡If one symbol represents 200 components, working correct ot the nearest⎤
⎣100 components gives: Mon. 8, Tues. 11, Wed. 9, Thur. 12 and Fri. $6\frac{1}{2}$.⎦

4  Draw a horizontal bar chart presenting the data given in *Problem 1* above.

⎡5 equally spaced horizontal rectangles, the lengths being proportional⎤
⎣to 41, 73, 62, 58 and 84 units respectively.⎦

5  For the data given in *Problem 2* above, draw a horizontal bar chart.

⎡6 equally spaced horizontal rectangles whose lengths are proportional⎤
⎣to 35, 44, 62, 68, 49 and 41 respectively.⎦

167

6 Present the data given in *Problem 3* above on a horizontal bar chart.

> ⌈5 equally spaced horizontal rectangles, whose lengths are proportional⌉
> ⌊to 1 580, 2 190, 1 840, 2 385 and 1 280 units respectively.       ⌋

7 Draw a vertical bar chart depicting the data given in *Problem 1* above.

> ⌈5 equally spaced vertical rectangles, whose heights are proportional⌉
> ⌊to 41, 73, 62, 58 and 84 units respectively.          ⌋

8 For the data given in *Problem 2* above, construct a vertical bar chart.

> ⌈6 equally spaced vertical rectangles, whose heights are proportional⌉
> ⌊to 35, 44, 62, 68, 49 and 41 units respectively.      ⌋

9 Depict the data given in *Problem 3* above on a vertical bar chart.

> ⌈5 equally spaced vertical rectangles, whose heights are proportional⌉
> ⌊to 1 580, 2 190, 1 840, 2 385 and 1 280 units respectively.    ⌋

10 A factory produces three different types of components. The percentages of each of these components produced for three, one-monthly periods are as shown below. Show this information on percentage component bar charts and comment on the changing trend in the percentages of the types of component produced.

| | Month 1 | 2 | 3 |
|---|---|---|---|
| Component P | 20 | 35 | 40 |
| Component Q | 45 | 40 | 35 |
| Component R | 35 | 25 | 25 |

> ⌈Three rectangles of equal height, sub-divided in the percentages⌉
> ⌊shown in the columns above. P increases by 20% at the expense of Q and R.⌋

11 An analysis of the expenditure of a small company in thousands of pounds for a two-year period is as shown below.

| | Year 1 | Year 2 |
|---|---|---|
| Salaries | 40 | 50 |
| Wages | 50 | 65 |
| Materials | 30 | 60 |
| Rent and rates | 15 | 25 |

Present this data on percentage component bar charts.

> ⌈Two rectangles of equal height, sub-divided into the following percentages:⌉
> ⌊year 1 30%, 37%, 22%, 11%; year 2 25%, $32\frac{1}{2}$%, 30%, $12\frac{1}{2}$%     ⌋

12 A company has five distribution centres and the mass of goods in tonnes sent to each centre during four, one-week periods, is as shown.

| | Week 1 | 2 | 3 | 4 |
|---|---|---|---|---|
| Centre A | 147 | 160 | 174 | 158 |
| Centre B | 54 | 63 | 77 | 69 |
| Centre C | 283 | 251 | 237 | 211 |
| Centre D | 97 | 104 | 117 | 144 |
| Centre E | 224 | 218 | 203 | 194 |

Use a percentage component bar chart to present this data and comment on any trends.

> ⌈Four rectangles of equal heights, sub-divided as follows:            ⌉
> | week 1 18%, 7%, 35%, 12%, 28%; week 2 20%, 8%, 32%, 13%, 27%;
> | week 3 22%, 10%, 29%, 14%, 25%; week 4 20%, 9%, 27%, 19% and 25%.
> | Little change in centres A and B, a reduction of about 5% in C, an
> ⌊increase of about 7% in D and a reduction of about 3% in E.         ⌋

13 A factory produces four different types of product. The value of each product produced in a day is as shown.

| Product | A | B | C | D |
|---|---|---|---|---|
| Value of Product (£) | 17 000 | 31 000 | 44 000 | 3 500 |

Present this data on a pie diagram.

[A circle of any radius, sub-divided into sectors having angles of 64°, 117°, 166° and 13° respectively.]

14 The employees in a company can be split into the following categories:
managerial 3, supervisory 9, craftsmen 21, semi-skilled 67, others 44.
Show this data on a pie diagram.

[A circle of any radius, sub-divided into sectors having angles of 7½°, 22½°, 52½°, 167½° and 110° respectively.]

15 The way in which an apprentice spent his time over a one-month period is as follows:

drawing office 44 hours, production 64 hours, training 12 hours, at college 28 hours.

Use a pie diagram to depict this information.

[A circle of any radius, sub-divided into sectors having angles of 107° 156°, 29° and 68° respectively.]

16 Form a frequency distribution having about 8 classes for the data given below, which refers to the diameter in millimetres of 60 bolts produced by an automatic process. Draw a histogram and a frequency polygon to depict this data.

```
3.97   4.02   4.05   4.00   3.98   3.99   4.04   3.97   3.99   3.95
4.01   3.99   4.01   3.96   4.03   4.01   3.99   4.01   4.02   4.04
3.95   4.06   3.98   4.02   4.00   3.97   3.96   4.00   4.07   4.00
3.99   4.02   4.03   3.99   4.01   4.03   4.02   4.01   4.01   4.02
3.98   4.00   3.93   4.05   4.00   3.99   4.00   3.98   4.06   3.97
4.08   4.04   4.01   3.97   3.95   4.02   3.99   4.03   3.99   4.00
```

[There is no unique solution, but one solution is:
3.92–3.94  1, 3.95–3.96  5, 3.97–3.98  9, 3.99–4.00  17, 4.01–4.02 15,
4.03–4.04  7, 4.05–4.06  4, 4.07–4.08  2.]

17 The information given below refers to the value of resistance in ohms of a batch of 48 resistors of similar value. Form a frequency distribution for the data, having about 6 classes and draw a frequency polygon and histogram to represent this data diagrammatically.

```
21.0   22.4   22.8   21.5   22.6   21.1   21.6   22.3
22.9   20.5   21.8   22.2   21.0   21.7   22.5   20.7
23.2   22.9   21.7   21.4   22.1   22.2   22.3   21.3
22.1   21.8   22.0   22.7   21.7   21.9   21.1   22.6
21.4   22.4   22.3   20.9   22.8   21.2   22.7   21.6
22.2   21.6   21.3   22.1   21.5   22.0   23.4   21.2
```

[There is no unique solution, but one solution is:
20.5–20.9  3, 21.0–21.4  10, 21.5–21.9  11,
22.0–22.4  13, 22.5–22.9  9, 23.0–23.4  2.]

18 The number of hours of overtime worked by a group of craftsmen during each of 48 working weeks in a year are as shown below. For this data (a) form a frequency distribution having about seven classes, and then (b) represent the grouped data diagrammatically by means of (i) a frequency polygon, and (ii) a histogram.

| 40 | 39 | 41 | 34 | 37 | 36 | 26 | 47 |
|----|----|----|----|----|----|----|----|
| 29 | 45 | 41 | 51 | 42 | 46 | 43 | 32 |
| 48 | 40 | 28 | 52 | 47 | 33 | 52 | 44 |
| 38 | 53 | 47 | 42 | 51 | 48 | 37 | 30 |
| 54 | 46 | 35 | 50 | 59 | 43 | 49 | 48 |
| 41 | 27 | 45 | 46 | 42 | 53 | 25 | 38 |

$$\begin{bmatrix} \text{There is no unique solution, but one solution is:} \\ 25-29 \quad 5, \ 30-34 \quad 4, \ 35-39 \quad 7, \ 40-44 \quad 11, \\ 45-49 \quad 12, \ 50-54 \quad 8, \ 55-59 \quad 1 \end{bmatrix}$$

19 (a) With reference to *Fig 5*, determine the amount spent on labour and materials to produce 1 650 units of the product.
   (b) With reference to *Fig 7(b)*, determine the total frequency between 7.65 and 8.55.
   (c) If in year 2 of *Fig 4*, 1% corresponds to 2.5 dwellings, how many bungalows are sold in that year?

$$[(a) \ \pounds495, \ (b) \ 34, \ (c) \ 88]$$

20 (a) If the company sell 23 500 units of the product depicted in *Fig 5*, determine the cost of their overheads per annum.
   (b) Using the histogram shown in *Fig 7(b)*, determine the total frequency being depicted.
   (c) If 1% of the dwellings represented in year 1 of *Fig 4* corresponds to 2 dwellings, find the total number of houses sold in that year.

$$[(a) \ \pounds16 \ 450, \ (b) \ 50, \ (c) \ 138]$$

21 Form a cumulative frequency distribution and hence draw the ogive for the frequency distribution given in the solution to *Problem 17*

$$\begin{bmatrix} 20.95 \quad 3, \ 21.45 \quad 13, \ 21.95 \quad 24, \\ 22.45 \quad 37, \ 22.95 \quad 46, \ 23.45 \quad 48 \end{bmatrix}$$

22 For the data given in the solution to *Problem 18*, form a cumulative frequency distribution and hence draw an ogive depicting this data.

$$\begin{bmatrix} 29.5 \quad 5, \ 34.5 \quad 9, \ 39.5 \quad 16, \\ 44.5 \quad 27, \ 49.5 \quad 39, \ 54.5 \quad 47, \\ 59.5 \quad 48 \end{bmatrix}$$

# Index